Robust Energy Procurement of Large Electricity Consumers

Sayyad Nojavan • Mahdi Shafieezadeh
Noradin Ghadimi
Editors

Robust Energy Procurement of Large Electricity Consumers

Springer

Editors
Sayyad Nojavan
Department of Electrical Engineering
University of Bonab
Bonab, Iran

Mahdi Shafieezadeh
Sharif University of Technology
Knowledge Management Office
Tehran, Iran

Noradin Ghadimi
Young Researchers and Elite Club
Ardabil Branch, Islamic Azad University
Ardabil, Iran

ISBN 978-3-030-03228-9 ISBN 978-3-030-03229-6 (eBook)
https://doi.org/10.1007/978-3-030-03229-6

Library of Congress Control Number: 2018964101

This Springer imprint is published by the registered company Springer Nature Switzerland AG
The registered company address is: Gewerbestrasse 11, 6330 Cham, Switzerland

Preface

In a restructured power market, large consumers are one of the most important players due to their high energy demand, which affects the economic and environmental performance of the power system. Additionally, obtaining energy at minimum cost will impart a considerable benefit for the large consumer. To this purpose, different options are available, namely, participating in the power market, self-generating units, and bilateral contracts. The required load demand of the consumer should be supplied while utilizing the abovementioned sources to incur a minimal power procurement cost. In the power procumbent process, different uncertainties affect the total cost of power procurement. It is well known that the power price in the pool market is uncertain, and is considered a main uncertainty source. The bilateral contracts, which are signed between the supplier and consumer before the physical delivery with predetermined prices and periods, are the most important tool used to cope with power price uncertainty.

Self-generating units can be used to reduce power procumbent sources. These days, different renewable energy sources are considered as an environment-friendly source to generate clean energy. Therefore, wind turbines and photovoltaic systems beside micro-turbines can be classified as self-generating units. Through implementing renewable energy sources, which are increasing daily, environmental goals can be satisfied; however, such energy sources create different uncertainties relating to the power procurement. For example, the power output of the photovoltaic system depends on the solar irradiation, and wind speed affects the power output of the wind turbine.

In order to deal with the aforementioned uncertainties, different methods are introduced to model the impact of the uncertainty. The most important methods are, specifically, stochastic programming, robust optimization, information-gap decision theory (IGDT), and novel hybrid methods; each has its own advantages and disadvantages. For example, the stochastic programming method, which is a scenario-based approach, is used only if there is adequate information about the uncertain parameter. The robust optimization only models the pernicious aspects of the uncertainty, while the IGDT models both pernicious and useful aspects. The latter methods, robust optimization and IGDT, model the impact of only one parameter; however, this problem is not an issue in the stochastic programming. Finally, hybrid methods, such as hybrid robust-stochastic or IGDT-stochastic, which are obtained by combining the abovementioned methods, can be considered to overcome the disadvantages and utilize the advantages of the chosen methods.

In order to reduce the procumbent cost, demand response programs (DRPs) can be utilized. DRPs transfer load from the peak periods to the off-peak period, which decreases the total power procurement cost. It is obvious that by implementing DRPs, the construction of new power plants to handle the peak load demand, is eliminated.

Besides the DRPs, which reduce the peak load demand, the electrical storage can be used to store energy during the low price and demand periods, which can then be used during the high price and demand periods; this reduces the generation of other power sources during peak demand.

Depending on what the large consumer is manufacturing, different energies, such as electricity, cooling, and heating, are required. The efficiency of such systems can be increased by utilizing new technologies, such as combined heat and power units (CHP); for example, the wasted heat energy generated can be used to supply the large consumer's heating requirement.

As previously stated, the high energy demand by a large consumer has a significant impact on environmental issues. To do so, a multi-objective model should be developed to deal with cost and emission. Different methods, such as weighted sum, goal programming, and compromise programming, are introduced to solve the multi-objective model.

The power procurement of a large consumer is examined in the 12 chapters that comprise this book. In the first chapter, the role of different power procurement sources is discussed. The problems surrounding the delivery of energy is solved in the deterministic case in Chap. 2. The stochastic programming, robust optimization, and IGDT methods are used to address the same issues in Chaps. 3, 4, and 5. The hybrid robust-stochastic and IGDT-stochastic methods are developed and implemented in Chaps. 6 and 7. The impact of time of use and real-time pricing demand response methods are studied in Chaps. 8 and 9. An energy hub model is introduced to solve the electricity and heating energy requirement of the large consumer in Chap. 10. In Chap. 11, the heating, cooling, and electricity energy demand of the large consumer is met by developing an energy hub model. Finally, in Chap. 12, the environmental effects of energy procurement by a large consumer are addressed by developing a multi-objective programming framework.

Bonab, Iran
Tehran, Iran
Ardabil, Iran

Sayyad Nojavan
Mahdi Shafieezadeh
Noradin Ghadimi

Contents

Contributors

Oveis Abedinia Department of Electrical Engineering, Budapest University of Technology and Economics, Budapest, Hungary

Adel Akbarimajd Department of Electrical Engineering, Faculty of Technical Engineering, University of Mohaghegh Ardabili, Ardabil, Iran

Raouf Morsali Asl Young Researchers and Elite Club, Ardabil Branch, Islamic Azad University, Ardabil, Iran

Sajjad Dadfar Department of Electrical Engineering, College of Electrical Engineering and Computer, Saveh Branch, Islamic Azad University, Saveh, Iran

Ministry of Energy, Iran Power Generation Transmission and Distribution Management Organization-Head Office (Tavanir), Tehran, Iran

Moeid Dehghanpour Farashah 3 Med Group, Tehran, Iran

Noradin Ghadimi Young Researchers and Elite Club, Ardabil Branch, Islamic Azad University, Ardabil, Iran

Mohammad Javad Golkar Department of Electrical and Computer Engineering, Zahedan Branch, Islamic Azad University, Zahedan, Iran

Mehrdad Khaksar Young Researchers and Elite Club, Islamshahr Branch, Islamic Azad University, Islamshahr, Iran

Hossein Khodaei Department of Engineering, Ardabil Branch, Islamic Azad University, Ardabil, Iran

Mojtaba Madadkhani Department of Information and Communications Technology, Malek Ashtar University of Technology, Tehran, Iran

Mahdi Mir Department of Electrical Engineering, Ferdowsi University of Mashhad, Mashhad, Iran

Sayyad Nojavan Department of Electrical Engineering, University of Bonab, Bonab, Iran

Alireza Rezvani Young Researchers and Elite Club, Saveh Branch, Islamic Azad University, Saveh, Iran

Iran Water and Power Resources Development Company (IWPCO), Tehran, Iran

Mahdi Shafieezadeh Sharif University of Technology, Knowledge Management Office, Tehran, Iran

Sayed Ahmad Reza Shokrani Department of Industrial Management, Faculty of Management, University of Tehran, Tehran, Iran

Milad Teimourian Sama Technical and Vocation Training College, Islamic Azad University, Parsabad Moghan Branch, Parsabad, Iran

Young Researchers and Elite Club, Germi Branch, Islamic Azad University, Ardabil, Iran

Ehsan Zargin Young Researchers and Elite Club, Ardabil Branch, Islamic Azad University, Ardabil, Iran

Chapter 1
The Concept of Large Consumer

Milad Teimourian, Noradin Ghadimi, Sayyad Nojavan, and Oveis Abedinia

Abstract In the restructured power system, a large consumer can provide its required demand by using multiple options. Pool market, bilateral contracts, and self-generating units are some of the available options for power procurement. The main goal of large consumers from participation in the power market is procuring power at the minimum cost. The power price in the pool market has uncertainty, and to cope with this problem, bilateral contracts with predetermined prices can be considered by a large consumer. To model uncertainty in the pool market, different methods such as stochastic programming, robust optimization approach, and information gap decision method can be used. Renewable energy sources can be used as self-generating units to meet some part of the required demand of the large consumer. Also, energy storage systems can be implemented to store energy during off-peak demand periods and use the stored energy during peak demand periods. Different types of storage systems such as electrical storage systems, thermal storage systems, and cooling storage systems are used to store electricity, heating, and cooling energies, respectively. In this chapter, reviews of considered methods to solve the problem are presented and the obtained results are discussed.

M. Teimourian
Sama Technical and Vocation Training College, Islamic Azad University, Parsabad Moghan Branch, Parsabad, Iran

Young Researchers and Elite Club, Germi Branch, Islamic Azad University, Ardabil, Iran

N. Ghadimi (✉)
Young Researchers and Elite Club, Ardabil Branch, Islamic Azad University, Ardabil, Iran
e-mail: noradin.ghadimi@iauardabil.ac.ir

S. Nojavan
Department of Electrical Engineering, University of Bonab, Bonab, Iran
e-mail: sayyad.nojavan@bonabu.ac.ir

O. Abedinia
Department of Electrical Engineering, Budapest University of Technology and Economics, Budapest, Hungary
e-mail: oveis.abedinia@vet.bme.hu

S. Nojavan et al. (eds.), *Robust Energy Procurement of Large Electricity Consumers*,
https://doi.org/10.1007/978-3-030-03229-6_1

1.1 Introduction

Restructuring in the electrical industry has started since the 1980s to create competition at all levels, from generation to consumer. This process has resulted in improvements of the planning and operational efficiencies of the power system. To achieve these goals, different models are proposed, and during the time, advantages and disadvantage of each model appeared. Figure 1.1 shows the competitive market model, which is accepted and applied by different countries and markets.

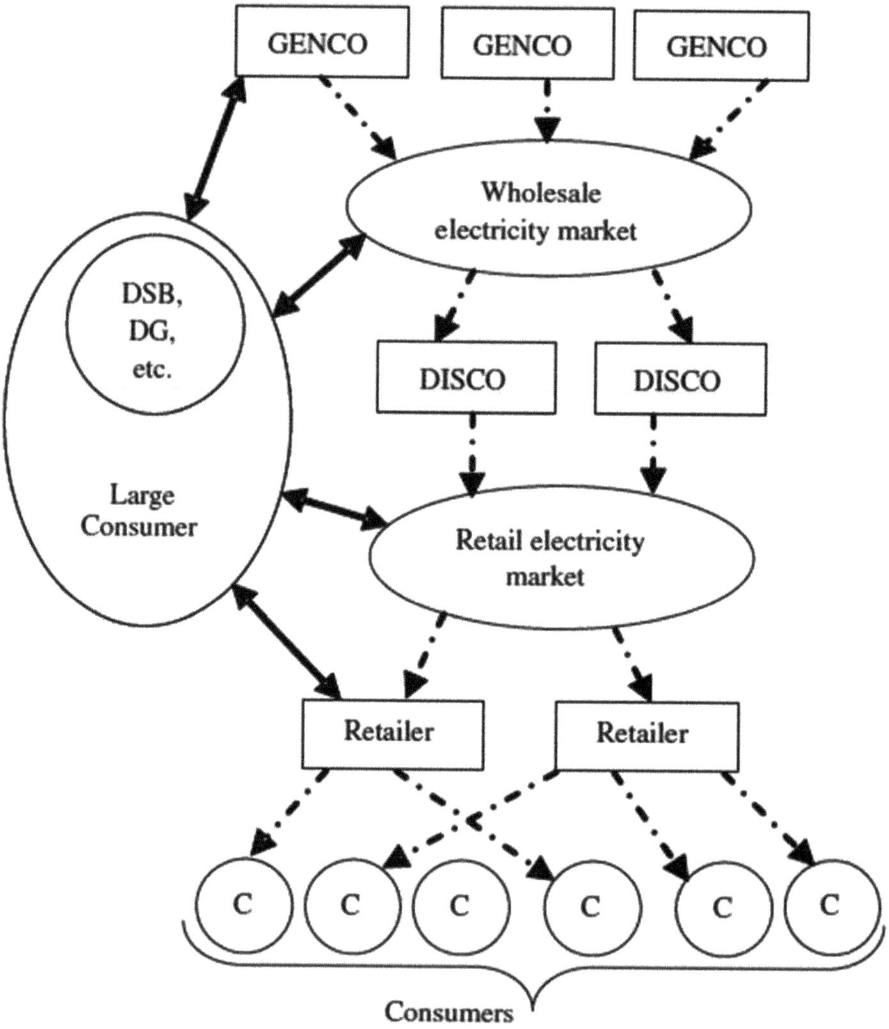

Fig. 1.1 Competitive market model

In the restructured power market, procuring power at the lowest possible cost is the main purpose of large consumers from taking part in the power market. To do so, each consumer has to select the best strategy to participate in the power market based on the market structure and its required demand.

Power procurement at the minimum cost has more importance for large consumers due to their high-required demand. A large consumer can be vehicle-assembling facilities, petrochemical industries, or aluminum production complexes. As shown in Fig. 1.1, in spite of other consumers, the large consumers have multiple sources to meet their demand. A large consumer can consider pool market, bilateral contracts, and self-generating units as the power procurement sources. Power price in the pool market has uncertainty, which may result in high procurement cost for the consumer. To reduce related risk to power price in the pool market, the consumer can sign bilateral contracts. Bilateral contracts are arrangements between a consumer and supplier, which allows the consumer to buy electrical energy before its physical delivery.

Besides the pool market and bilateral contracts, to reduce risk and cost, the consumer can use self-generating units to meet some part of its demand. Different types of distributed energy resources (DERs) can be considered as self-generating units. Renewable energy sources as a DER can be taken into account by considering climate characteristics of the location of the consumer. In addition, in some countries, the consumer can benefit government incentives by using renewable energy sources. To overcome volatile output of renewable energy sources, electrical storage systems (ESS) can be utilized which store electrical energy when power generation is high and use the stored energy when the renewable energy source is not available.

By using demand response programs, a consumer can manage its load during peak and off-peak times, which will result in reducing energy procurement cost. Demand response programs transfer some part of the load from peak time to off-peak periods. Among introduced demand response programs, time-varying programs are more suitable for the assumed problem. These types of programs are divided into three main groups as time-of-use pricing (TOU), critical peak pricing/rebates (CPP/CPR), and real-time pricing (RTP).

In the TOU, by assigning higher prices during peak periods, power prices are changed over the hours of the day. CPP/CPR increases prices or provides rebates for conservation during the critical peak hours. Finally, RTP allows prices to vary dynamically by considering the marginal electricity cost.

As said before, the consumer can utilize the abovementioned sources to meet its demand at the lowest possible price. Figure 1.2 shows a schematic diagram of the proposed model for power procurement of a large consumer.

1.1.1 Literature Review

Power procurement problem of a large consumer can be assayed from different point of views. Besides electrical energy, a large consumer can require heating and cooling energies. In this case, published papers on the topic are divided into three main

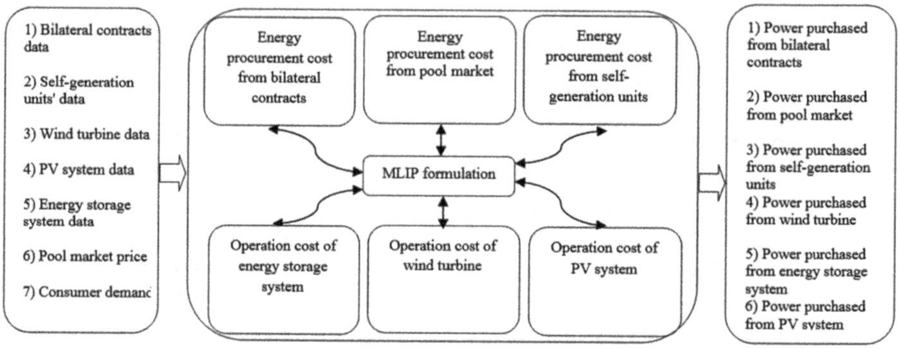

Fig. 1.2 Schematic diagram of power procurement sources of a large consumer

groups as electricity, electricity and heat, and electricity, heating, and cooling. In the following subsection, analyses of published studies are presented.

1.1.1.1 Electricity-Based Large Consumer

Various resources can be used to solve the energy procurement of a large electricity consumer. In [1], electricity procurement of a large consumer is studied using pool market, bilateral contracts, and self-generation units. Renewable energy sources such as photovoltaic (PV) and wind turbine (WT) are taken as self-generating units. To model the pool price uncertainty, information decision gap theory (IGDT) is applied. By using 12 bilateral contracts, procurement risk is reduced, which comes from power price uncertainty. Demand response programs are used to reduce procurement cost by adjusting load based on power price. Obtained results show using DRP has reduced about 8% of the procurement cost in all of IGDT strategies.

Stochastic programming-based technique is introduced in [2] to solve the power procurement problem considering bilateral contracts, a limited amount of self-production, and the pool. Minimizing procurement cost while limiting the cost fluctuation risk is investigated in [3]. A procedure to optimally decide purchasing a strategy from different electricity sources is provided in [4] considering a decision horizon spanning from 1 to several months.

Obtained results in [5] show that the related strategies to procurement with a higher cost are more risk-averse and robust against pool market price uncertainty. In addition, the obtained results show that the participation of sources with uncertainty has decreased when the power price has increased. It is also shown that buying strategy is a risk-seeking strategy when the procurement costs are lower than the expected minimum cost. This issue may provide unexpected benefits.

As a numerical example, the robustness against power price uncertainty in procurement decision will be about 9.5%, by assigning $8,500,000 to meet electricity demand. This amount is 5.5% higher than the minimum procurement cost.

A novel approach based on the IGDT method is introduced in [6] to get the optimal bidding curve by considering the same problem and using the same case study in [5]. The bidding curves are derived for the same case study by considering multi-markets (day-ahead market and a subsequent adjustment market) in [7].

The power procurement problem is solved in [8] by deriving optimal offering strategy which is robust against power price fluctuations. In this Ref., the impact of time-of-use DRP is considered on the process. In addition, power price uncertainty is modeled using IGDT method, although the proposed method in this Ref. does not minimize the procurement cost but allows taking robust decision against price uncertainty in the market. To solve power procurement problem, an auction model is developed in [9] which can be used to determine electricity price between the large consumers and competitive generators by considering the transaction and transmission costs in a wholesale electricity market.

Bidding curve of an hour for generators and large consumers in the pool market is derived by using a teaching-learning-based optimization method in [10]. The result of this paper is compared with other heuristic methods to show the nobility of the proposed method. The obtained results for generators, in comparison with the previous work of the author, show that the profit of generators is reduced.

A new fuzzy-based decision-making system is proposed in [11] for procurement of electricity using different resources to get compromise between the profit and risk. The fuzzy α-cuts are used to model the price uncertainty in the problem to make the decision for maximizing profit of the large consumer considering different α-cuts corresponding to different uncertainty levels. Depending on being a risk-taker or being risk-averse, the consumer can choose the appropriate strategy among the provided range of different decision-making strategies of profit uncertainty.

In order to get optimal scheduling of large industrial consumers, a new decision-making framework is proposed in [12]. Pool market, bilateral contracts, and internal generation have been considered as power sources. Power price and the rate of availability of the generating unit are considered as main sources of uncertainty. The problem is solved using stochastic programming and tried to minimize procured power from the pool market. Three bilateral contracts are considered, namely, peak, shoulder, and valley periods of a day. Obtained results have shown that by reducing purchasing power from the pool market, the reliability of the system has increased. Finally, it should be denoted that the proposed framework is presented for mid-time scheduling.

Power procurement problem is solved in [13] for a 1-year time horizon considering bilateral contracts and spot market. Stochastic programming is used to model uncertainty of demand. In this paper, bilateral contracts are divided into seasonal and annual contracts in which seasonal contracts apply time-of-use demand response program while the annual contract has a determined price. Note that the objective function in this work has concentrated on the total power procurement cost.

In [14], a novel decision-making method based on IGDT is proposed for a large consumer. Pool market, self-generating unit, and 12 bilateral contracts are main sources of power procurement. The proposed method in this study helps decision-maker to get the better strategy to cope with power price uncertainty. Obtained results show that to be robust against power price volatility, procured power from pool market should be minimized, and bilateral contracts with predetermined prices should be put on priority.

In [15], ESS and renewable energy sources are used to meet large consumer demand. In addition, in order to reduce power procurement cost, demand response programs are applied. By using DRP, the procurement costs of the consumer are reduced because the load is shifted from high-price periods to low-price periods, which in this case the load profile of the consumer is flattened and the maximum demand in each period is decreased. In this study, 3.4% reduction in the cost of power procurement is obtained. Finally, the authors in [16] have derived optimal bidding curves for each hour for the assumed case study in [15].

A summary of reviewed papers is provided in Table 1.1.

Table 1.1 Summary of published articles on power procurement problem of a large consumer

	Power source				Uncertainty		Renewable energy				
Ref.	PM	BC	SF	Method	Price	Load	PV	WT	DRP	ESS	Bidding curve
[1]	✓	✓	✓	IGDT	✓		✓	✓		✓	
[2]	✓	✓		Stochastic programming	✓						
[3]	✓	✓	✓	Variance of the cost	✓						
[5]	✓	✓	✓	IGDT							
[6]	✓	✓	✓	IGDT	✓		✓	✓	✓	✓	
[7]	✓			IGDT	✓						
[8]	✓			Stochastic programming	✓						✓
[10]	✓			Stochastic programming	✓						✓
[11]	✓	✓	✓	Probabilistic fuzzy	✓						
[12]	✓	✓	✓	Stochastic programming	✓						
[13]	✓	✓		Stochastic programming		✓					
[14]	✓	✓	✓	IGDT	✓						
[15]	✓	✓	✓	Stochastic programming	✓	✓	✓	✓	✓	✓	
[16]	✓	✓	✓	Stochastic programming	✓	✓	✓	✓	✓	✓	✓

PM pool market, *BC* bilateral contracts, *SF* self-generating unit, *PV* photovoltaic system, *WT* wind turbine, *DRP* demand response program, *ESS* electric storage system

1.1.1.2 Electricity- and Heat-Based Large Consumer

Electricity is not the only required energy by the large consumers. In some industries, heating energy is essential in the manufacturing process. Some part of heating energy can be obtained by using combined heat and power (CHP) generation units. In CHP units, wasted heat in the power-producing process can be used to meet heat load demand. Besides the CHP units, combination of heating and power hubs can be utilized to meet heat and power demand. Figure 1.3 shows the sample model which is used in [17].

1.1.1.3 Electricity-, Heating-, and Cooling-Based Large Consumers

Three kinds of demand for heating, cooling, and electrical energy can be required in the manufacturing or processing of a large industrial consumer. These demands can be supplied simultaneously by using combined cooling, heating, and power (CCHP) units, which are composed of different energy suppliers such as gas turbine, gas boiler, and electric and absorption chiller [18]. By combining these energy suppliers, the total efficiency of the system will increase because some suppliers use wasted energy of other units.

The CCHP-based energy hub system in the presence of renewable energies such as WT and PV is implemented to simultaneously improve efficiency and reduce the operation cost and emission in [19]. Obtained results indicate that the maximum improvement in the exergy efficiency and product cost is obtained within 33.49% using the LINMAP method and 23.67% using the TOPSIS procedure. On the other hand, the highest improvement in the environmental impact rate is calculated to be about 22.7% with 0.01788.

Detailed information about these systems is provided in Chap. 11.

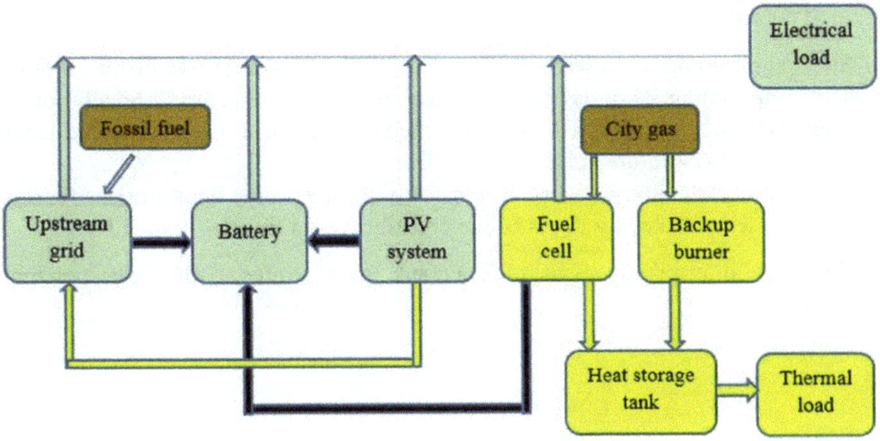

Fig. 1.3 Schematic diagram of the heating and power hub model

1.2 Price Forecasting Methods

As said in the previous sections, market price volatility imposes risks on the power procurement problem, which may cause a high procurement cost. By considering this issue, a good forecast of electricity power price in the short term is very important for a large consumer to participate in the pool market by submitting buying bids to the market operator. In a pool-based electric energy market, producers submit selling bids containing energy blocks and their corresponding minimum selling prices to the market operator. Similarly, consumers submit buying bids containing energy blocks and their corresponding maximum buying prices to the market operator. Then, the market operator clears the market using an appropriate market-clearing procedure, which results in accepted selling and buying bids of hourly energy prices. In this situation, forecasting price by consumers and producers is crucial to get optimum selling and bidding curves which results in profit maximization of producers and utility maximization for consumers. In spite of demand series, nonconstant mean and variance characteristics of price series have made electricity price forecasting very difficult.

Several techniques have been introduced to forecast short-term electricity prices, which are divided into two groups as hard and soft computing techniques.

Autoregressive integrated moving average (ARIMA) [20], time series model [21], mixed model [22], and wavelet-ARIMA [23] are the most important hard computing techniques.

Hybrid intelligent system (HIS) [24], cascaded neuro-evolutionary algorithm (CNEA) [25], neuro-fuzzy models [26], neural networks (NN) [27], weighted nearest neighbors (WNN) [28], fuzzy neural networks (FNN) [29], and adaptive wavelet neural network (AWNN) [30] are the most important soft computing techniques.

Each of the abovementioned methods uses a special procedure to forecast electricity power price in the short-term period. For example, the wavelet-ARIMA method considers the following steps to forecast 24 hourly prices of a day:

- Step (1): Decompose through the wavelet transform the available historical price series (up to hour 24 of day (d-1)) in a set of constitutive series (typically four).
- Step (2): Use a specific ARIMA model fitted to each one of the constitutive series to forecast its 24 future values for day d.
- Step (3): Use the inverse wavelet transform to forecast the hourly prices for day d using the estimates for day d of the constitutive series.

For more information about each technic, please refer to the corresponding reference.

1.3 Demand Response Programs

Deregulation and restructuring of the electricity industry are carried out to increase the efficiency of the power system by keeping differences between peak and low load periods as small as possible [31]. For a reliable operation of the electricity system, perfect balance between the demand and supply is necessary in the real-time considerations. On the other hand, the participation of the customer in the wholesale energy markets tends to pay lower average electricity prices and reduce the price volatility. Demand response program is defined as changes in electric consumption patterns of end-user clients in response to changes of electricity price over time or to incentive payments designed to decrease high electricity usage at high wholesale market price times or when the system reliability problems occur [32]. In other words, the demand response (DR) is the procedure that consumers respond to the price signals inserted in tariffs by changing their consumption patterns [33]. In addition, to reduce the price volatility during peak demand hours, the independent system operator (ISO) can utilize DR programs.

To reduce the price spikes, a new framework has been proposed using demand response program in [34]. By using successful DR program, the demand over the day can be smoothed out without decreasing energy consumption, which results in avoiding the construction of expensive power plants to serve the peaks that occur for just a few hours per year. This issue can be considered as the most important advantage of DR programs. On the other hand, DR offers the required flexibility at relatively low cost. The result of DR investigations and execution in US utilities and power markets is discussed in [35].

Different demand response programs (DRP) can be classified into two main categories: incentive-based programs (IBPs) and price-based programs (PBPs). More details of these groups can be found in [35, 36]. Figure 1.4 illustrates the classification of demand response programs.

PBPs are based on the dynamic pricing rates in which the electricity tariffs are not flat. PBPs tend to flatten the demand curve by offering lower prices during off-peak periods and high price during peak periods. The time-of-use (TOU) rate, critical peak pricing (CPP), extreme day pricing (EDP), extreme day CPP (ED-CPP), and real-time pricing (RTP) are the most important PBP rates [34, 35]. The TOU rates are the basic type of PBP, which are the rates of electricity price per unit consumption that differ in different blocks of time [34, 35]. The peak and off-peak are the simplest TOU rates. The rate design tries to reflect the average cost of electricity during different periods. Designing process of a TOU rate is well studied in [37, 38]. The CPP rates include a pre-specified higher electricity usage price superimposed on TOU rates or normal flat rates. CPP prices are used during high wholesale electricity prices or contingencies for a limited number of hours or days per year [39]. According to [40], EDP differs from CPP in the fact that the price is in effect for the whole 24 h of the extreme day, which is unknown until a day ahead. On the other hand, EDP is similar to CPP in having a higher price for electricity.

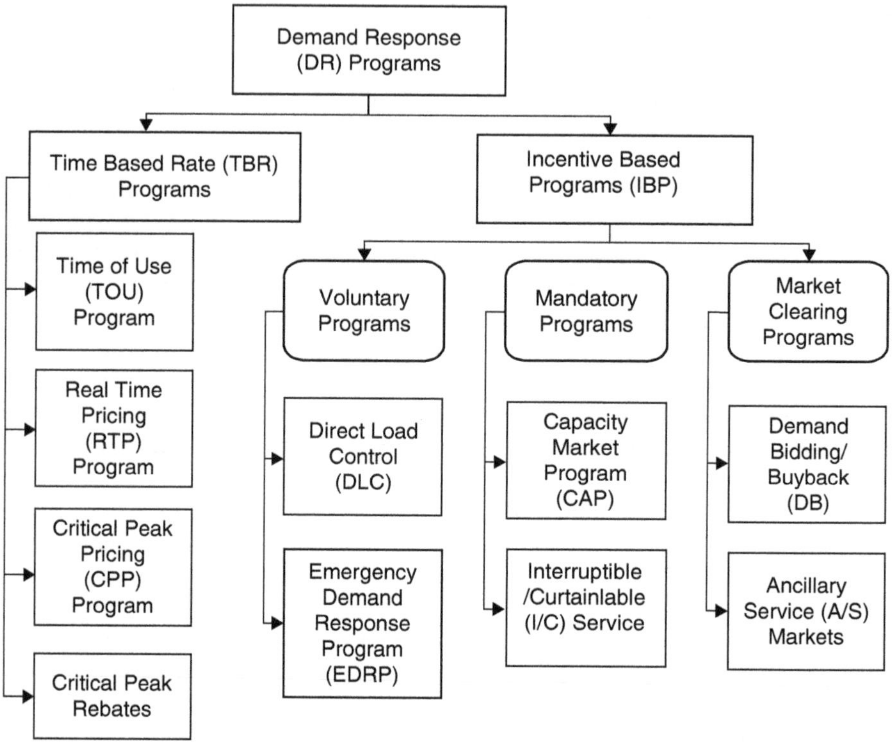

Fig. 1.4 Classification of demand response programs

In RTPs, the customers are charged with hourly fluctuating prices, which reflect the real electricity cost in the market. RTP customers are informed about the prices on a day-ahead or hour-ahead basis. RTP programs are introduced as the most efficient and direct DR programs which are suitable for competitive electricity markets [41].

PBPs have spanned a wide range of planning horizon from a few minutes to many years. In this case, DRP can be categorized based on their planning intervals into long-term, mid-term, and short-term programs. The long-term or mid-term programs consider more than a day period. Therefore, to solve the energy procurement problem of a large consumer with a mid-term self-scheduling, the TOU and RTP programs can be considered. The impacts of TOU and RTP programs on the power procurement problem of a large consumer are studied in Chaps. 8 and 9, respectively.

1.4 Renewable Energy

As discussed in Sect. 1.1, self-generation units can be taken as a power procurement source for a large consumer. By considering environmental issues, different types of renewable energy sources can be considered as a distributed energy resource. Wind turbine, PV, fuel cell (FC), etc. can be selected based on climate characteristics of the location of the large consumer. Although using renewable energy resources will increase uncertainty in the system due to their volatile nature, to cope with environmental issues such as climate change and considering fossil fuel prices, using renewable energy resources is inevitable.

In the following subsection, brief information about the modeling of WT and PV systems are presented.

1.4.1 *Wind Turbine Model*

To produce electrical power, the kinetic energy of the wind can be converted to the electrical energy through wind turbines. To investigate the uncertainty of wind speed, the Weibull distribution is considered to model the uncertainty of the wind speed. In this way, by considering the Weibull distribution curve of wind speed, discrete scenarios are generated [42]. Then, the maximum produced power by the wind turbine in each time t and scenario s can be calculated using Eq. (1.1) [43].

$$P_{t,s}^{\text{wind, max}} = \begin{cases} 0 & V_{t,s}^{w} < V_{\text{ci}} \\ P_r \times \left(\frac{V_{t,s}^{w} - V_{\text{ci}}}{V_r - V_{\text{ci}}}\right)^3 & V_{\text{ci}} < V_{t,s}^{w} < V_{\text{cr}} \\ P_r & V_r < V_{t,s}^{w} < V_{c0} \\ 0 & V_{t,s}^{w} > V_{c0} \end{cases} \tag{1.1}$$

It should be mentioned that the energy procured by a large electricity consumer from WT in each time t and scenario s should be limited as shown in Eq. (1.2).

$$P_{t,s}^{\text{wind}} \leq P_{t,s}^{\text{wind, max}}; \quad \forall t, s \tag{1.2}$$

1.4.2 *Photovoltaic System Model*

Solar energy can directly be converted into electricity by PV system. In order to get maximum power output of the system, the maximum power point tracking (MPPT) can be implemented in the PV systems [44]. By using MPPT systems, the maximum power output of the system can be obtained for all radiations and temperatures.

The forecasted error distribution curves can be used to model the uncertainty of radiations and temperature. These curves are divided into some intervals with the width of one standard deviation. Therefore, considering the normal distribution curve of radiations and temperature, the scenarios of PV system are generated. According to [45], the maximum available output power of PV systems can be calculated for each scenario s and time t using Eq. (1.3).

$$P_{t,s}^{M,\max} = \frac{G_{t,s}^{a}}{G_{a_0}} \times \left\{ P_{\mathrm{Max},0}^{M} + \mu_{P\max} \times \left(T_{t,s}^{a} + G_{t,s}^{a} \times \frac{\mathrm{NOCT} - 20}{800} - T_{M,0} \right) \right\} \tag{1.3}$$

Procured power by the large consumer should be limited as shown in Eq. (1.4) for each scenario s and time t.

$$P_{t,s}^{\mathrm{PV}} \le P_{t,s}^{M,\max}; \quad \forall t, s \tag{1.4}$$

1.5 Storage Systems

Storage systems can be used to store energy during off-peak periods and use stored energy during peak periods. Generally, peak energy prices occur during peak demand periods. This means that by using the discussed procedure, economic benefits can be experienced [46]. For example, storing electric power during low-price periods, which is experienced during off-peak periods, and selling stored energy during high-price periods can make a profit to the owner of the storage system. Progress in technology has made storing different kinds of energies possible. Thermal storage systems (TSS) and cooling storage systems (CSS) can be used to store thermal and cooling energies, respectively.

As said before, the large or industrial consumer may require heating and cooling energies besides electricity power in the manufacturing process. In this case, using electric storage systems (ESSs), TSS, and CSS can be considered by the consumer, which may result in lower power procurement cost.

1.5.1 Electrical Storage System

It is obvious that most of the electrical energy is consumed during the daytime than nighttime, because of the socioeconomic conditions. This leads to an increase in the peak-valley difference of day and night in the power demand. To cope with the increase in the peak-valley difference, new power plants should be set up. During this condition, thermal power units and hydropower stations are operated at maximum possible capacities.

On the other hand, both the electrical demand and generation are fluctuating in the power system. The characteristics of end-user consumption determine demand fluctuation, and renewable energy sources are one of the primary sources of fluctuation in the generation side. The stability of the grid will be affected by frequent demand fluctuation.

To get the balance between load and generation, EES during excess generation hours can act as a load. Reversely, it can act as a generator, during peak demand hours. As said before, EES can contribute to lower the power purchase cost during peak demand hours.

To cope with sudden load increase, EES can take part by providing instant energy to meet unscheduled required demand. On the other hand, this issue will reduce power procurement from pool market in this particular case, which may reduce power procurement cost considering the uncertain market price.

1.5.2 Thermal Storage System

For different thermal applications, thermal energy storage (TES) is a promising technology with a high potential. This is a very attractive technology because it is considered as the most appropriate solution to correct the gap between the demand and supply of energy.

There are three ways as thermochemical heat storage (THS), sensible heat storage (SHS), and latent heat storage (LHS). Among these storage techniques, THS is the most promising alternative to be used as an energy storage system. To generate heat, THS systems utilize both chemical and sorption reactions. For more information about the structure of the TSS systems, please refer to [47].

1.5.3 Cooling Storage System

In order to reduce the electricity peak power demand of air conditioning systems and operating costs in the industrial process, the large industrial consumers can utilize the cooling storage systems. Moreover, some CSSs can offer higher system efficiency and lower first costs in comparison with the non-storage systems. Without requiring any additional power generation equipment, CSS can shift the power load of air conditioning systems from daytime to nighttime.

CSSs generally include ice thermal storage (ITS) system, eutectic salt cool thermal storage (ESCTS) system, and chilled water thermal storage (CWTS) system. Detailed information about each of the abovementioned systems can be found in [48].

The CSS system operates based upon three types of conventional control strategies, which determine the operating modes. Figure 1.5 shows these strategies. For more information about these strategies, please refer to [48].

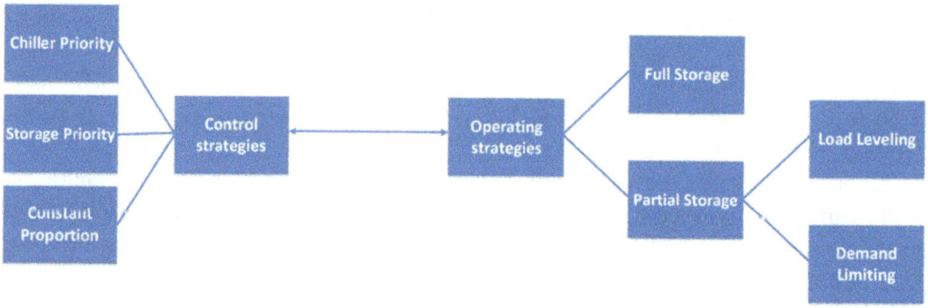

Fig. 1.5 Control strategies of cooling storage systems

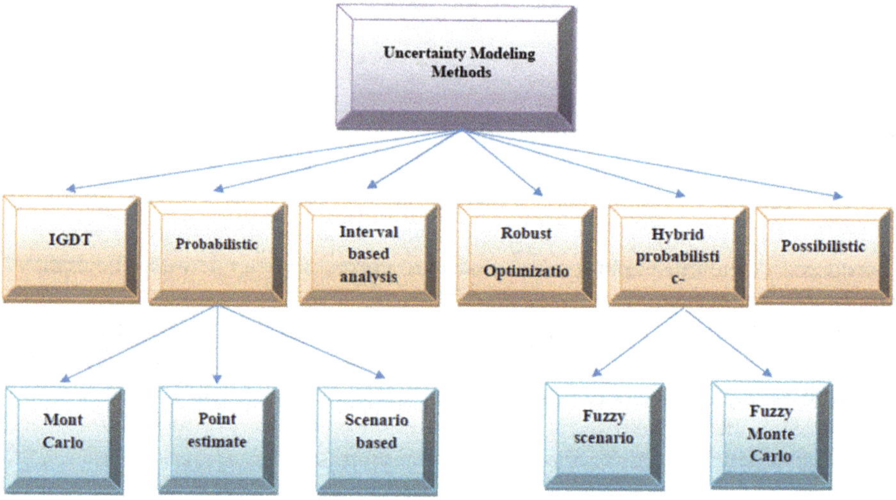

Fig. 1.6 Uncertainty modeling methods

1.6 Uncertainty Modeling

As said before, the power price in the pool market has uncertainty. In order to get better decisions in the presence of this uncertainty, using uncertainty modeling is inevitable. Different methods have been proposed for this purpose. Figure 1.6 represents the classification of the uncertainty modeling methods.

The most important and useful methods are stochastic programming, robust optimization, and information gap decision theory. Each of these methods has their advantage and disadvantages. For example, stochastic programming suffers from the high computational burden and needs scenario reduction methods [16]. In the following subsections, each of the abovementioned methods is briefly introduced.

1.6.1 Uncertainty Modeling Using Monte Carlo Simulation

To model different uncertainty in a system, the Monte Carlo simulation (MCS) method, which is a numerical simulation procedure, can be applied. This method considers random variables with known or assumed probability distributions. A deterministic simulation process is repeated, where according to the corresponding probability distributions a particular set of values for the random variables are generated in each simulation. The results of the Monte Carlo simulation can be considered as a sample of an experimental observation. To apply the methods of statistical estimation and inference to the data set obtained, the results of many such simulations should be collected.

For the power procurement problem of the large consumer, the random variable to be treated by the simulation procedure is the power price considering its stochastic behavior. It is assumed that the electricity power price variable is normally distributed with a known mean value (considering the forecasted value) and a known standard deviation. To get appropriate values of the random variable, inverting the cumulative distribution function can be used. In particular, to get normally distributed random numbers directly, the MATLAB function *randn* is an option.

1.6.2 Uncertainty Modeling Using Two-Point Estimate Method

The Monte Carlo simulation is an iterative approach utilizing cumulative density function (CDF) of random variables to get the result. Obtaining the desired convergence by considering the great number of iterations is the main disadvantage of the MCS. Unlike the MCS, the two-point estimate method is an approximate method. In this method, some representative points (s points for each variable), which are called concentrations, should be determined using the provided information by central moments. Obtained representative points are used for solving the model, and the statistical information of the random output variable is calculated using the obtained solutions [49].

Assume $X\{x_1, x_2, \ldots, x_l, \ldots, x_m\}$ as a random variable with a mean value μ_{x_l} and standard deviation σ_{x_l} and Z as a random function of X (i.e., $Z = \text{F}(X)$). Each of the s concentrations of variables x_l is defined as a pair, including a weight $w_{l,s}$ and location $x_{l,s}$. Using Hong's two-point estimate method, function F has to be evaluated only s times for each random input variable x_l at the points made up of the s.th location of random input variable x_l and the mean value (μ_{x_l}) of the remaining input variables. Therefore, $2 \times m$evaluations should be carried out. By using Eq. (1.5), the location $x_{l,s}$ is calculated [50].

$$x_{l,s} = \mu_{x_l} + \xi_{l,s} \cdot \sigma_{x_l} \tag{1.5}$$

in which $\xi_{l,s}$ is the standard location of random variable x_l. The standard locations and weights of random variable of x_l are computed by

$$\xi_{l,1} = \frac{\lambda_{l,3}}{2} + \sqrt{m + \left(\frac{\lambda_{l,3}}{2}\right)^2}, \quad \xi_{l,2} = \frac{\lambda_{l,3}}{2} - \sqrt{m + \left(\frac{\lambda_{l,3}}{2}\right)^2} \tag{1.6}$$

and

$$w_{l,1} = -\frac{\xi_{l,2}}{m(\xi_{l,1} - \xi_{l,2})}, \quad w_{l,2} = \frac{\xi_{l,1}}{m(\xi_{l,1} - \xi_{l,2})} \tag{1.7}$$

where $\lambda_{l,3}$ denotes the skewness of the random variable x_l

$$w_{l,1} = -\frac{\xi_{l,2}}{m(\xi_{l,1} - \xi_{l,2})}, \quad w_{l,2} = \frac{\xi_{l,1}}{m(\xi_{l,1} - \xi_{l,2})} \tag{1.8}$$

The algorithm for solving the power procurement problem using Hong's two-point estimate method is shown in Fig. 1.7.

In the power procurement problem, the power price is modeled as a random variable with a known probability distribution. The weights and locations should be calculated based on the abovementioned procedure. For each concentration, a deterministic power procurement should be run. The solution of power procurement problem is

$$Z_{l,s} = F\{x_{l,1}, x_{l,2}, \ldots, x_{l,s}, \ldots, x_{m,s}\} \tag{1.9}$$

where $Z_{l,s}$ is the vector of random output variables associated with the s.th concentration of random input variable and represents the nonlinear relation between the input and output variables in the problem. By using Eq. (1.10), the raw moments of output random variables are calculated.

$$E(Z) \cong E(Z) + \sum_{s} w_{l,s} \cdot Z_{l,s} \tag{1.10}$$

Flowchart of the two-point estimate method is summarized in the following steps:

- Step 1: Set the first and second moments of s.th output random variables to zero: $E(Z) = 0$.
- Step 2: Select the input random variable x_l.
- Step 3: Compute $\lambda_{l,3}$,$\xi_{l,s}$,$w_{l,s}$ using Eqs. (1.5)–(1.7).
- Step 4: Determine the two estimated locations $x_{l,s}$.

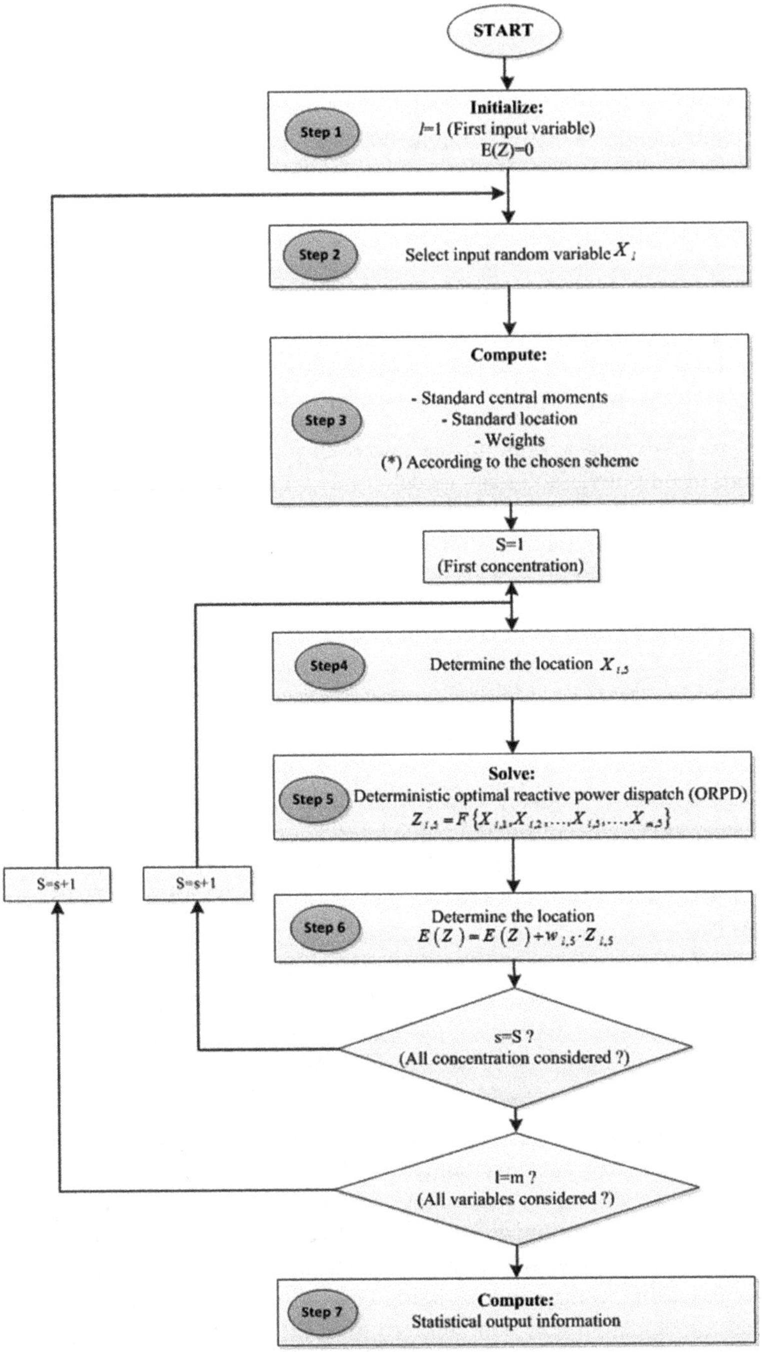

Fig. 1.7 Flowchart of the two-point estimate algorithm

- Step 5: Solve the deterministic power procurement problem for each concentration.
- Step 6: Update the raw moments of output variables.
- Step 7: Repeat steps 2–6 until all concentrations of all input random variables are taken into account. Finally, compute the statistical information of output random variables.

1.6.3 Uncertainty Modeling Using Interval Optimization-Based Model

To model uncertainty using interval optimization method, a known interval is assumed for assigning a value to the uncertain parameter. By using this method, the bounds of output variables can be determined. Equations (1.11)–(1.13) present a standard formulation for the optimization model subjected to equal and unequal constraints considering the uncertain parameter ρ.

$$\text{Min} f(X, U, \rho) \tag{1.11}$$

$$\begin{aligned} &s.t. \\ &g(X, U, \rho) = 0 \end{aligned} \tag{1.12}$$

$$h(X, U, \rho) \leq 0 \tag{1.13}$$

As said before, instead of expected values, upper and lower values of the uncertain parameter are considered in the interval optimization approach. Therefore, a lower and upper bound should be calculated for the objective function. To calculate the lower and upper bounds of objective function, Eqs. (1.14) and (1.15) are used, respectively.

$$f^{-}(X) = \min_{\rho \in U} f(X) \tag{1.14}$$

$$f^{+}(X) = \max_{\rho \in U} f(X) \tag{1.15}$$

To input interval parameter as an uncertain parameter, interval objective function is created. Therefore, an uncertainty-based interval objective function should be minimized. To do so, an uncertainty-based interval objective function can be transformed into a deterministic multi-objective model involving average power procurement cost and deviation cost, which are the conflict objective functions.

In this case, the average and deviation costs should be minimized. It should be noted that the deviation cost is minimized until the strategy of the large consumer can be more robust against market price uncertainty. Equations (1.16)–(1.18) present the multi-objective model based on interval optimization approach.

$$\text{Min} f(X) = \text{Min}\left(f^M(X), f^W(X)\right) \tag{1.16}$$

$$f^M(X) = \frac{f^+(X) + f^-(X)}{2} \tag{1.17}$$

$$f^W(X) = \frac{f^+(X) - f^-(X)}{2} \tag{1.18}$$

It should be mentioned that $f^M(X)$ and $f^W(X)$ are average cost and deviation cost of large industrial consumer.

1.6.4 Stochastic Programming

The stochastic programming models the uncertainty by using discrete scenarios, and the decisions are made at different stages. By using PDF, scenario generation can be carried out. Considering the high number of scenarios will increase computational burden and time. To cope with this problem, scenario reduction methods can be used in which these methods reduce the number of scenarios. One of the most important and useful methods of scenario reduction is the Kantorovich distance methods. A detailed description of the proposed method is presented in [15]. More detail about the stochastic programming method will be presented in Chap. 3.

1.6.5 Uncertainty Modeling Using the Robust Optimization Method

As said before, stochastic methods are less preferred due to the high information requirement about uncertain parameters. The robust optimization approach is another method to handle uncertainty parameters, which can be considered as a substitution for stochastic programming to address uncertainty in mathematical programming models. The robust optimization approach is a risk management method that has a low computing volume in comparison with other methods [51]. This method investigates the effect of an uncertain parameter on the optimal result, which aims to reduce the sensitivity of the optimal result to the uncertain parameter.

More information about this method is presented in Chap. 4.

1.6.6 Uncertainty Modeling Using Information Gap Decision Theory

One of the most important methods to model uncertainty in the market price is the information gap decision theory (IGDT). In spite of other uncertainty modeling methods, additional information about the probability density function of the uncertain parameter is not required in this method. In addition, IGDT provides accurate and efficient results in comparison with stochastic methods in which their results are dependent on probabilistic scenarios [37]. The advantages of IGDT method over other methods are summarized as follows:

1. Additional information such as the probability density function of stochastic variables is not required.
2. The accurate and efficient results in comparison with stochastic programming methods.

The decision-maker can use IGDT method in order to evaluate risks and recognize priorities and opportunities which will lead toward getting a better decision. Uncertainties may have negative or positive effects. For example, the negative effect can lead to high cost, while positive effect can lead to an unexpected profit. In order to consider the abovementioned issues, two functions are proposed in IGDT as opportunity and robustness.

The robustness function investigates the harmful or negative aspect of uncertainty. In the assumed problem, the robustness function is the degree of resistance to uncertainty against high procurement cost.

The opportunity function assays the useful or positive aspect of uncertain parameter and analyzes the possibility of getting benefits.

The decision-maker has to make a decision based on obtained quantitation from the robustness and opportunity functions. In addition, the decision-maker will determine how much robustness against harmful uncertain parameter is acceptable. The same question can be considered for opportunity function. It should be denoted that there is no algorithm to find unique answers to these questions. The decision-maker should give the answer to these questions based on his/her experience. More information about this method is presented in Chap. 5.

1.7 Conclusion

In the restructured power market, large consumers can participate in the power market to get their own demand. Different power sources such as pool market, bilateral contracts, and self-generating are available for a large consumer. Pool market is the main power source for the large consumer, and power price in this market has uncertainty, which increases power procurement risks. To reduce risk, a large consumer can sing the bilateral contract with predetermined prices. On the other hand, self-generating units can be an option to meet some part of the required demand.

Procuring power at the minimum cost is the main motivation of large consumers to participate in the pool market. To get the better decision in the uncertain environment of the pool market, uncertainty modeling methods as stochastic programming, robust optimization, and information gap decision theory should be used.

In addition, using demand response programs is another option to reduce power procurement cost. These programs transfer some part of the load from the high-price period to low-price periods which results in lower procurement cost.

Based on climate characteristics of the location of the large consumer, renewable energy sources (RES) can be considered as self-generating units to meet some part of its demand and reduce emission. Wind turbine and photovoltaic system are the most prevailing RES to get electricity, although, according to RES volatile nature, uncertainties will be imposed to the system.

To cope with the abovementioned risks and uncertainties, energy storage systems can be used to reduce peak-valley difference of daytime and nighttime, which can help the economical operation of the system.

References

1. H.A. Bagal, Y.N. Soltanabad, M. Dadjuo, K. Wakil, N. Ghadimi, Risk-assessment of photovoltaic-wind-battery-grid based large industrial consumer using information gap decision theory. Sol. Energy **169**, 343–352 (2018)
2. M. Carrion, A.B. Philpott, A.J. Conejo, J.M. Arroyo, A stochastic programming approach to electric energy procurement for large consumers. IEEE Trans. Power Syst. **22**(2), 744–754 (2007)
3. A.J. Conejo, M. Carrion, Risk-constrained electricity procurement for a large consumer. IEE Proc. Gener. Transm. Distrib. **153**(4), 407 (2006)
4. A.J. Conejo, J.J. Fernandez-Gonzalez, N. Alguacil, Energy procurement for large consumers in electricity markets. IEE Proc. Gener. Transm. Distrib. **152**(3), 357 (2005)
5. K. Zare, M.P. Moghaddam, M.K. Sheikh El Eslami, Electricity procurement for large consumers based on information gap decision theory. Energy Policy **38**(1), 234–242 (2010)
6. S. Nojavan, H. Ghesmati, K. Zare, Robust optimal offering strategy of large consumer using IGDT considering demand response programs. Electr. Power Syst. Res. **130**, 46–58 (2016)
7. K. Zare, A.J. Conejo, M. Carrión, M.P. Moghaddam, Multi-market energy procurement for a large consumer using a risk-aversion procedure. Electr. Power Syst. Res. **80**(1), 63–70 (2010)
8. S.J. Kazempour, A.J. Conejo, C. Ruiz, Strategic bidding for a large consumer. IEEE Trans. Power Syst. **30**(2), 848–856 (2015)
9. D. Fang, J. Wu, D. Tang, A double auction model for competitive generators and large consumers considering power transmission cost. Int. J. Electr. Power Energy Syst. **43**(1), 880–888 (2012)
10. R.K. Mallick, R. Agrawal, P.K. Hota, Bidding strategies of Gencos and large consumers in competitive electricity market based on TLBO, in *2016 IEEE 6th International Conference on Power Systems (ICPS)*, (2016), pp. 1–6
11. M. Zarif, M.H. Javidi, M.S. Ghazizadeh, Self-scheduling approach for large consumers in competitive electricity markets based on a probabilistic fuzzy system. IET Gener. Transm. Distrib. **6**(1), 50 (2012)
12. M. Zarif, M.H. Javidi, M.S. Ghazizadeh, Self-scheduling of large consumers with second-order stochastic dominance constraints. IEEE Trans. Power Syst. **28**(1), 289–299 (2013)
13. Q. Zhang, A.M. Bremen, I.E. Grossmann, J.M. Pinto, Long-term electricity procurement for large industrial consumers under uncertainty. Ind. Eng. Chem. Res. **57**(9), 3333–3347 (2018)
14. K. Zare, M.P. Moghaddam, M.K. Sheikh-El-Eslami, Risk-based electricity procurement for large consumers. IEEE Trans. Power Syst. **26**(4), 1826–1835 (2011)

15. S. Nojavan, H. allah Aalami, Stochastic energy procurement of large electricity consumer considering photovoltaic, wind-turbine, micro-turbines, energy storage system in the presence of demand response program. Energy Convers. Manag. **103**, 1008–1018 (2015)
16. S. Nojavan, H. Qesmati, K. Zare, H. Seyyedi, Large consumer electricity acquisition considering time-of-use rates demand response programs. Arab. J. Sci. Eng. **39**(12), 8913–8923 (2014)
17. B.M. Soudmand, N.N. Esfetanaj, S. Mehdipour, R. Rezaeipour, Heating hub and power hub models for optimal performance of an industrial consumer. Energy Convers. Manag. **150**, 425–432 (2017)
18. X. Chen, G. Gong, Z. Wan, C. Zhang, Z. Tu, Performance study of a dual power source residential CCHP system based on PEMFC and PTSC. Energy Convers. Manag. **119**, 163–176 (2016)
19. F.A. Boyaghchi, M. Chavoshi, Monthly assessments of exergetic, economic and environmental criteria and optimization of a solar micro-CCHP based on DORC. Sol. Energy **166**, 351–370 (2018)
20. J. Contreras, R. Espinola, F.J. Nogales, A.J. Conejo, ARIMA models to predict next-day electricity prices. IEEE Trans. Power Syst. **18**(3), 1014–1020 (2003)
21. H. Zareipour, C.A. Cañizares, K. Bhattacharya, J. Thomson, Application of public-domain market information to forecast Ontario's wholesale electricity prices. IEEE Trans. Power Syst. **21**(4), 1707–1717 (2006)
22. C. Garcia-Martos, J. Rodriguez, M.J. Sanchez, Mixed models for short-run forecasting of electricity prices: application for the Spanish market. IEEE Trans. Power Syst. **22**(2), 544–552 (2007)
23. A.J. Conejo, M.A. Plazas, R. Espinola, A.B. Molina, Day-ahead electricity price forecasting using the wavelet transform and ARIMA models. IEEE Trans. Power Syst. **20**(2), 1035–1042 (2005)
24. N. Amjady, M. Hemmati, Day-ahead price forecasting of electricity markets by a hybrid intelligent system. Eur. T. Electr. Power **19**(1), 89–102 (2009)
25. N. Amjady, F. Keynia, Day-ahead price forecasting of electricity markets by mutual information technique and cascaded neuro-evolutionary algorithm. IEEE Trans. Power Syst. **24**(1), 306–318 (2009)
26. C.P. Rodriguez, G.J. Anders, Energy price forecasting in the Ontario competitive power system market. IEEE Trans. Power Syst. **19**(1), 366–374 (2004)
27. J.P.S. Catalão, S.J.P.S. Mariano, V.M.F. Mendes, L.A.F.M. Ferreira, Short-term electricity prices forecasting in a competitive market: a neural network approach. Electr. Power Syst. Res. **77**(10), 1297–1304 (2007)
28. A.T. Lora, J.M.R. Santos, A.G. Exposito, J.L.M. Ramos, J.C.R. Santos, Electricity market price forecasting based on weighted nearest neighbors techniques. IEEE Trans. Power Syst. **22**(3), 1294–1301 (2007)
29. N. Amjady, Day-ahead price forecasting of electricity markets by a new fuzzy neural network. IEEE Trans. Power Syst. **21**(2), 887–896 (2006)
30. N.M. Pindoriya, S.N. Singh, S.K. Singh, An adaptive wavelet neural network-based energy price forecasting in electricity markets. IEEE Trans. Power Syst. **23**(3), 1423–1432 (2008)
31. M.H. Albadi, E.F. El-Saadany, A summary of demand response in electricity markets. Electr. Power Syst. Res. **78**(11), 1989–1996 (2008)
32. H.A. Aalami, H. Pashaei-Didani, S. Nojavan, Deriving nonlinear models for incentive-based demand response programs. International Journal of Electrical Power & Energy Systems. 2019 Mar 1;106:223–31
33. R.S. Ferreira, L.A. Barroso, M.M. Carvalho, Demand response models with correlated price data: a robust optimization approach. Appl. Energy **96**, 133–149 (2012)
34. R. Aazami, K. Aflaki, M.R. Haghifam, A demand response based solution for LMP management in power markets. Int. J. Electr. Power Energy Syst. **33**(5), 1125–1132 (2011)

35. D. Kathan, R. Aldina, M.P. Lee, L. Medearis, P. Sporborg, M. Tita, D. Wight, A. Wilkerson, F. Kreger, V. Richardson, W. Gifford, Assessment of demand response and advanced metering. Rapport technique. Federal Energy Regulatory Commission, USA. 2012 Dec.
36. H.A. Aalami, M. Parsa Moghaddam, G.R. Yousefi, Modeling and prioritizing demand response programs in power markets. Electr. Power Syst. Res. **80**(4), 426–435 (2010)
37. Y. Tang, H. Song, F. Hu, Y. Zou, Investigation on TOU pricing principles, in *IEEE/PES Transmission and Distribution Conference and Exposition: Asia and Pacific*, (2005), pp. 1–9
38. C.W. Gellings, J.H. Chamberlin, *Demand-side Management: Concepts and Methods* (The Fairmont Press, Lilburn, 1987)
39. P. Jazayeri et al., A survey of load control programs for price and system stability. IEEE Trans. Power Syst. **20**(3), 1504–1509 (2005)
40. C. River, Primer on demand-side management with an emphasis on price-responsive programs. Prepared for The World Bank by Charles River Associates. Tech Rep. 2005
41. J. Edward, P. Policy, Assessment of customer response to real time pricing. New Jersey: Edward J, Bloustein School of Planning and Public Policy, State University of New Jersey. 2005 Jun 30
42. H. Leng, X. Li, J. Zhu, H. Tang, Z. Zhang, N. Ghadimi, A new wind power prediction method based on ridgelet transforms, hybrid feature selection and closed-loop forecasting. Adv. Eng. Inform **36**, 20–30 (2018)
43. M. Abbaspour, M. Satkin, B. Mohammadi-Ivatloo, F. Hoseinzadeh Lotfi, Y. Noorollahi, Optimal operation scheduling of wind power integrated with compressed air energy storage (CAES). Renew. Energy **51**, 53–59 (2013)
44. O. Abedinia, N. Amjady, N. Ghadimi, Solar energy forecasting based on hybrid neural network and improved metaheuristic algorithm. Comput. Intell. **34**(1), 241–260 (2018)
45. D.T. Nguyen, L.B. Le, Optimal bidding strategy for microgrids considering renewable energy and building thermal dynamics. IEEE Trans. Smart Grid **5**(4), 1608–1620 (2014)
46. F. Mirzapour, M. Lakzaei, G. Varamini, M. Teimourian, N. Ghadimi, A new prediction model of battery and wind-solar output in hybrid power system. J. Ambient. Intell. Humaniz. Comput., 1–11 (2017)
47. D. Aydin, S.P. Casey, S. Riffat, The latest advancements on thermochemical heat storage systems. Renew. Sust. Energ. Rev. **41**, 356–367 (2015)
48. X. Song, L. Liu, T. Zhu, S. Chen, Z. Cao, Study of economic feasibility of a compound cool thermal storage system combining chilled water storage and ice storage. Appl. Therm. Eng. **133**, 613–621 (2018)
49. J.M. Morales, J. Perez-Ruiz, Point estimate schemes to solve the probabilistic power flow. IEEE Trans. Power Syst. **22**(4), 1594–1601 (2007)
50. C.-L. Su, Probabilistic load-flow computation using point estimate method. IEEE Trans. Power Syst. **20**(4), 1843–1851 (2005)
51. S. Nojavan, A. Najafi-Ghalelou, M. Majidi, K. Zare, Optimal bidding and offering strategies of merchant compressed air energy storage in deregulated electricity market using robust optimization approach. Energy **142**, 250–257 (2018)

Chapter 2
Deterministic-Based Energy Procurement

Mahdi Shafieezadeh, Adel Akbarimajd, Noradin Ghadimi, and Mojtaba Madadkhani

Abstract In this chapter, the deterministic-based energy procurement problem of a large consumer is solved considering the six alternative power sources including bilateral contracts, pool market, wind turbine, photovoltaic system, self-generating units, and energy storage systems in the presence of the time-of-use rate of demand response program. The objective function and constraints are defined to minimize the total power procurement cost. The power price in the pool market is considered as predetermined values to get deterministic results. The time-of-use rate of demand response programs, as an effective way to reduce procurement cost, is implemented to decrease the total power procurement cost. In order to investigate the impact of demand response program, the problem is solved in two discrete case studies as without and with considering demand response program. The problem is formulated as a mixed-integer linear programming and solved by the CPLEX solver under GAMS optimization software. Obtained results in this section can be used as input to some uncertainty modeling methods to investigate the uncertainty of power price in the market, load demand, or any other uncertainty in the problem.

M. Shafieezadeh
Sharif University of Technology, Knowledge Management Office, Tehran, Iran

A. Akbarimajd
Department of Electrical Engineering, Faculty of Technical Engineering, University of Mohaghegh Ardabili, Ardabil, Iran
e-mail: akbarimajd@uma.ac.ir

N. Ghadimi (✉)
Young Researchers and Elite Club, Ardabil Branch, Islamic Azad University, Ardabil, Iran
e-mail: noradin.ghadimi@iauardabil.ac.ir

M. Madadkhani
Department of Information and Communications Technology, Malek Ashtar University of Technology, Tehran, Iran

S. Nojavan et al. (eds.), *Robust Energy Procurement of Large Electricity Consumers*,
https://doi.org/10.1007/978-3-030-03229-6_2

2.1 Introduction

As discussed in Chap. 1, a large consumer has multiple options to procure its demand in the restructured power system. Pool market (PM), bilateral contracts (BCs), and self-generation units are some of the available options. In addition, it is explained that the power price in the pool market, as the main power source for a large consumer, is an uncertain parameter. In this chapter, to discuss the determin istic method results, power price in the pool market is considered as a known parameter. In addition, it should be denoted that deterministic simulation results could be required by some uncertainty modeling methods [1]. For example, as explained in Chap. 1, in two-point estimate method, deterministic power procurement should be run for each concentration. In this way, power price in the pool market is considered as a predetermined value, and the problem is solved without considering any uncertainty related to power price [2]. Results are obtained for a case study consisting 12 bilateral contracts with different predetermined prices for different periods and self-generating units. In addition to bilateral contracts, different renewable energy sources are considered to supply the required load of the large consumer. Cost function of self-generating units is modeled using three piecewise linear models [3]. Renewable energy sources (RES) including photovoltaic (PV) system and wind turbine (WT) are considered as distributed energy resources (DER). To reduce power procurement cost, energy storage systems are assumed to store energy during low-price period and use stored energy to meet load demand in high-price periods. Finally, excessive required demand is procured from the pool market. In order to reduce power procurement cost of the large consumer, demand response programs (DRPs) can be taken into account as an effective and reasonable solution [4]. Among different introduced DRP models in Chap. 1, the time-of-use (TOU) rate of demand response programs is considered to reduce power procurement results.

The problem is solved by considering two different cases as without and with DRP to examine the impact of DRP.

2.2 Problem Formulation

Procuring the required power at the minimum possible cost is the main object of a large consumer from participating in the power market [5]. Therefore, the decision maker has to choose the best strategy by considering available sources. To do so, an objective function should be defined by considering operational constraint to minimize the power procurement cost [6]. In the following subsection, an objective function involving different power sources for energy procurement and related constraints is defined. Considered power sources include pool market (PM), bilateral contracts (BCs), photovoltaic system (PV), wind turbine (WT), self-generation units, and electrical storage systems (ESSs). The proposed model for solving the problem is shown in Fig. 2.1 considering demand response programs.

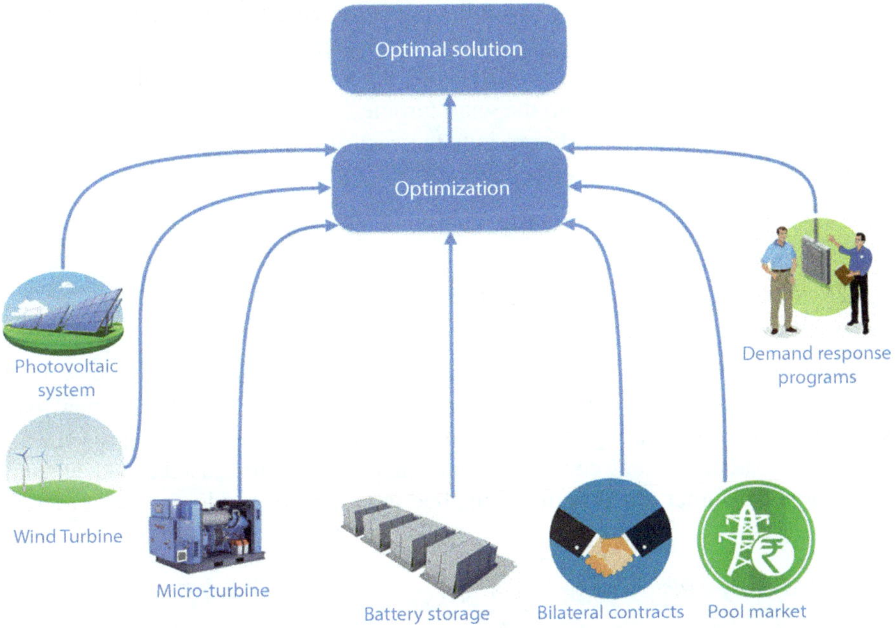

Fig. 2.1 Proposed case model to solve the power procurement problem of large consumer

2.2.1 Objective Function

Equation (2.1) presents the objective function of the power procurement problem of a large consumer.

$$\underset{P_t^p, P_{l,t}^{\mathrm{BC}}, P_{j,h,t}, P_t^{\mathrm{charge}}, P_t^{\mathrm{disc}}, P_t^{\mathrm{wind}}, P_t^{PV}}{\text{Minimize}} \sum_{l=1}^{B}\sum_{t=1}^{T}\lambda_{l,t}P_{l,t}^{\mathrm{BC}} + \sum_{t=1}^{T}\left\{\begin{array}{l} \lambda_t P_t^p + \sum_{j=1}^{N_j}\sum_{h=1}^{N_h} S_{j,h}^{\mathrm{DG}} P_{j,h,t} \\ + C_t^B \times \left(\chi \times P_t^{\mathrm{charge}} + \dfrac{P_t^{\mathrm{disc}}}{\eta}\right) \\ + C_t^{\mathrm{wind}} \times P_t^{\mathrm{wind}} + C_t^{\mathrm{PV}} \times P_t^{\mathrm{PV}} \end{array}\right\} \tag{2.1}$$

The first term of Eq. (2.1) presents the cost of procured power from the BCs $P_{l,t}^{\mathrm{BC}}$ with predetermined price $\lambda_{l,t}$ where t is the index of time (hour), l is the index of bilateral contracts, and B is the number of bilateral contracts. The cost of energy procurement from the PM P_t^p with the market price λ_t at time t is presented by the second term. The third term presents the cost of produced power by the jth unit of the self-generation units $P_{j,h,t}$ at time t with the related cost of block h of unit jth $S_{j,h}^{\mathrm{DG}}$.

The cost of using battery storage with the operating cost C_t^B at time t considering charging/discharging efficiency χ/η and charging/discharging power of battery storage (MW) $P_t^{\text{charge}}/P_t^{\text{disc}}$ at time t is modeled by the fourth term. The fifth term shows the cost of produced power from the wind turbine P_t^{wind} at time t with the operating cost C_t^{wind}. Finally, the last term presents the cost of produced power from the PV system P_t^{PV} at time t with operating cost C_t^{PV}.

2.2.2 Power Balance Constraint

Procured power from the alternative sources should meet the required new load considering DRP at time t, which is shown in constraint (2.2). Note that the new load is obtained by implementing the time-of-use rate demand response program (DRP) on the base load of the large consumer. Note that in order to forecast the load demand of the large consumer, the introduced method in [7] can be used.

$$\sum_{l=1}^{B} P_{l,t}^{\text{BC}} + P_t^p + \sum_{j=1}^{N_j} \sum_{h=1}^{N_h} P_{j,h,t} + P_t^{\text{wind}} + P_t^{\text{PV}} + P_t^{\text{disc}}$$

$$= \left((1 - \text{DR}_t) \cdot \text{load}_t^0 + \text{ldr}_t \right) + P_t^{\text{charge}} \tag{2.2}$$

where
DR_t is the participation factor of load in DRP
load_t^0 is the initial load at time t
ldr_t is the amount of shifted load from other load level to tth-hour load level

2.2.3 Constraints of Bilateral Contracts

Procured power from BCs should be in an acceptable interval which is described by Eq. (2.3). Total procured power from different BCs is summed in Eq. (2.4).

$$P_{l,t}^{\min} s_l < P_{l,t}^{\text{BC}} < P_{l,t}^{\max} s_l; \quad \forall l, t \tag{2.3}$$

$$P_t^{\text{BC}} = \sum_{l=1}^{N_l} P_{l,t}^{\text{BC}}; \quad \forall t \tag{2.4}$$

where
$P_{l,t}^{\min}$ is the minimum capacity relating to contracts l at time t
$P_{l,t}^{\max}$ is the maximum capacity relating to contracts l at time t
s_l is the binary variable, "1" if bilateral agreement l is selected, and 0 otherwise
P_t^{BC} is the total procured power from the bilateral contracts at time t

2.2.4 Model of Self-Generation Units

Operating cost of self-generation (SG) units is modeled using three piecewise linear curves which is shown in Fig. 2.2. Equation (2.5) shows the operating cost of self-generation units. Operational constraints of these units are shown in constraints (2.6)–(2.11) in which constraints (2.6) and (2.7) state that the sum of procured energy by the SG units should be less than the maximum capacity of the three blocks. The ramping up and down rate limits are described by constraints (2.8) and (2.9), respectively. Finally, constraint (2.10) expresses the minimum up and down time constraints by using auxiliary variables $\mathrm{Up}_{i,j}$ and $\mathrm{Dn}_{i,j}$ which are defined by constraint (2.11).

$$\mathrm{Cost} = \sum_{t=1}^{T}\sum_{j=1}^{N_j}\sum_{h=1}^{N_h} S_{j,h}^{\mathrm{DG}} P_{j,h,t} \tag{2.5}$$

$$0 \leq P_{j,h,t} \leq \left(P_{j,h}^{\mathrm{MAX}} - P_{j,h-1}^{\mathrm{MAX}}\right) \times U_{j,t};\quad \forall j,h,t \tag{2.6}$$

$$0 \leq P_{j,1,t} \leq P_{j,1}^{\mathrm{MAX}} \times U_{j,t};\quad \forall j,t \tag{2.7}$$

$$\sum_{h=1}^{N_h} P_{j,h,t} - \sum_{h=1}^{N_h} P_{j,h,t-1} \leq R_j^{\mathrm{up}} \times U_{j,t};\quad \forall j,t \tag{2.8}$$

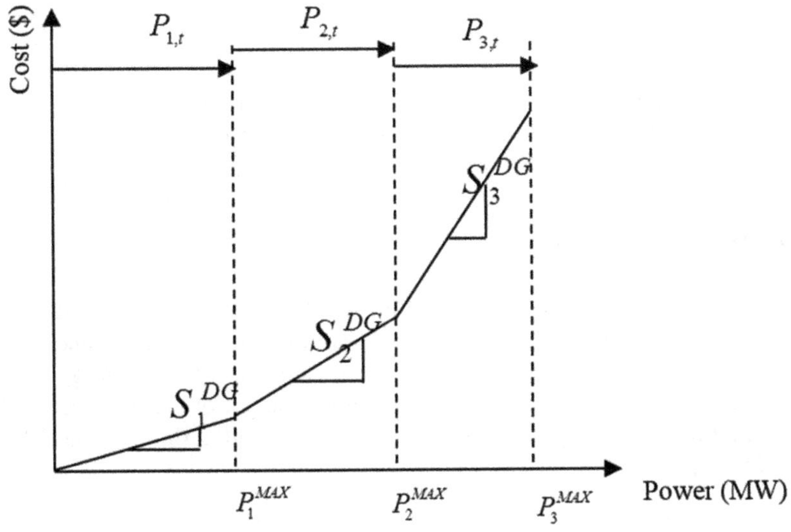

Fig. 2.2 Operation cost model of self-generation units

$$\sum_{h=1}^{N_h} P_{j,h,t-1} - \sum_{h=1}^{N_h} P_{j,h,t} \leq R_j^{\text{down}} \times U_{j,t-1}; \quad \forall j, t \tag{2.9}$$

$$\begin{aligned} U_{j,t} - U_{j,t-1} &\leq U_{j,t+\text{Up}_{j,i}}; \quad \forall j, \forall t, \forall i \\ U_{j,t-1} - U_{j,t} &\leq 1 - U_{j,t+\text{Dn}_{j,i}}; \quad \forall j, \forall t, \forall i \end{aligned} \tag{2.10}$$

$$\begin{aligned} \text{Up}_{j,i} &= \begin{Bmatrix} i & i \leq \text{MUT}_j \\ 0 & i \succ \text{MUT}_j \end{Bmatrix} \\ \text{Dn}_{j,i} &= \begin{Bmatrix} i & i \leq \text{MDT}_j \\ 0 & i \succ \text{MDT}_j \end{Bmatrix} \end{aligned} \tag{2.11}$$

where

$P_{j,h}^{\text{MAX}}$ is the power output of block h of jth unit of SG units [MWh]

$U_{j,t}$ is the binary variable equal to one if SG j unit is on mode at time t and 0 otherwise

$R_j^{\text{up}}, R_j^{\text{down}}$ is the ramp-up/ramp-down rate limit of j unit of SG [MW/h]

MUT_j,MDT_j is the minimum up/down time of self-generation units j [h]

2.2.5 *Renewable Energy-Based Power Sources*

By using RESs, greenhouse emissions to the atmosphere can be reduced which can help to cope with environmental issues [8]. Furthermore, the role of fossil fuel sources, as the main source of energy with volatile prices, in a power-producing process can be decreased which may result in lower power procurement cost [9]. It should be denoted that penetration of RESs increases day by day all around the word and meanwhile researchers try to reduce the operating cost of using these resources [10]. In this way, WT and PV, which are the most promising renewable energy sources (RESs), are considered as DER to supply the large consumer [11].

2.2.5.1 Wind Turbine Model

As said in Chap. 1, renewable energy can be used to meet some part of the required load of a large consumer [12]. In order to harness wind energy, wind turbines (WT) can be used to generate electric power from the kinetic energy of wind [13]. In this chapter, as discussed before, the power output of the WT is considered as a deterministic value [14]. Procured power of wind turbine is modeled by Eq. (2.12) [15].

$$P_{t,s}^{\text{wind,max}} = \begin{cases} 0 & V_{t,s}^{w} < V_{\text{ci}} \\ p_r \times \left(\frac{V_{t,s}^{w} - V_{\text{ci}}}{V_r - V_{\text{ci}}}\right)^3 & V_{\text{ci}} < V_{t,s}^{w} < V_{cr} \\ p_r & V_r < V_{t,s}^{w} < V_{c0} \\ 0 & V_{t,s}^{w} > V_{c0} \end{cases} \tag{2.12}$$

where

V_t^w is the wind speed at time t
V_r is the rated speed [m/s]
V_{ci} is the cut-in speed [m/s]
V_{c0} is the cutout wind speed [m/s]

Procured energy from WT at each time t is limited by using Eq. (2.13).

$$P_t^{\text{wind}} \leq P_t^{\text{wind,max}}; \quad \forall t \tag{2.13}$$

2.2.5.2 Photovoltaic System Model

Photovoltaic (PV) systems are another renewable energy sources, which can be used by a large consumer [16]. PV systems are modeled using Eq. (2.14). As said in Chap. 1, in order to get maximum power from PV systems, it is assumed that the maximum power point tracking (MPPT) system is always used for operation of the PV system [17]. More information about the proposed PV model is provided in Chap. 1.

$$P_t^{M,\max} = \frac{G_t^a}{G_{a_0}} \times \left\{ P_{\text{Max},0}^{M} + \mu_{P\max} \times \left(T_{t,}^a + G_t^a \times \frac{\text{NOCT} - 20}{800} - T_{M,0} \right) \right\} \tag{2.14}$$

where

$P_t^{M,\max}$ is the maximum power at any temperature T_t^a
G_t^a is the insulation for each time t
G_{a_0} is the insulation
$P_{\text{Max},0}^{M}$ is the maximum power
$T_{M,0}$ is the module temperature at the standard condition
NOCT is the normal operating cell

Procured power form PV systems are limited using Eq. (2.15).

$$P_t^{\text{PV}} \leq P_t^{M,\max}; \quad \forall t \tag{2.15}$$

2.2.6 Energy Storage System Model

Energy storage systems (ESSs) are used to reduce power procurement of the large consumers. ESS is modeled using Eqs. (2.16)–(2.21) [18]. Initial energy of the ESS is presented by Eq. (2.16). Constraints (2.17) and (2.18) limit the charging and discharging power at time *t*, respectively. Furthermore, the constraint (2.19) presents the level of stored energy in the ESS. Simultaneously charging and discharging of the ESS are prevented by constraint (2.20). Finally, the energy dynamic model is described by constraint (2.21) [19].

$$X_{t_0}^b = X_0^b \tag{2.16}$$

$$P_t^{\text{charge}} \leq P_{\text{charge}}^{\max} \times U_t^{\text{charge}}; \quad \forall t \tag{2.17}$$

$$P_t^{\text{disc}} \leq P_{\text{disc}}^{\max} \times U_t^{\text{disc}}; \quad \forall t \tag{2.18}$$

$$X_b^{\min} \leq X_t^b \leq X_b^{\max}; \quad \forall t \tag{2.19}$$

$$U_t^{\text{charge}} + U_t^{\text{disc}} \leq 1; \quad \forall t \tag{2.20}$$

$$X_t^b = X_{t-1}^b + \chi \times P_t^{\text{charge}} - \frac{P_t^{\text{disc}}}{\eta}; \quad \forall t \tag{2.21}$$

where

X_t^b is the stored energy in the ESS at time *t*
$P_t^{\text{charge}}/P_t^{\text{disc}}$ is the charging/discharging power of ESS (MW) at time
$P_{\text{charge}}^{\max}/P_{\text{disc}}^{\max}$ is the maximum charging/discharging power at time *t*
$U_t^{\text{charge}}/U_t^{\text{disc}}$ is the binary variable, which is equal to one if charging/discharging of the ESS at time *t*
$X_b^{\max}, X_b^{\min}$ is the maximum/minimum stored energy in the ESS

2.2.7 Demand Response Program Model

The importance of demand response programs is discussed in Chap. 1. To reduce the energy procurement cost, DRP transfers load from high-price periods to low-price periods [20]. According to [21], a large consumer can implement the time-of-use (TOU) rate of DRP, which has a limit on the maximum shifting capability of the load. This limit is assumed 20% in this chapter. Figure 2.3 illustrates the TOU rate of DRP.

Equations (2.22) and (2.23) mathematically model Fig. 2.3.

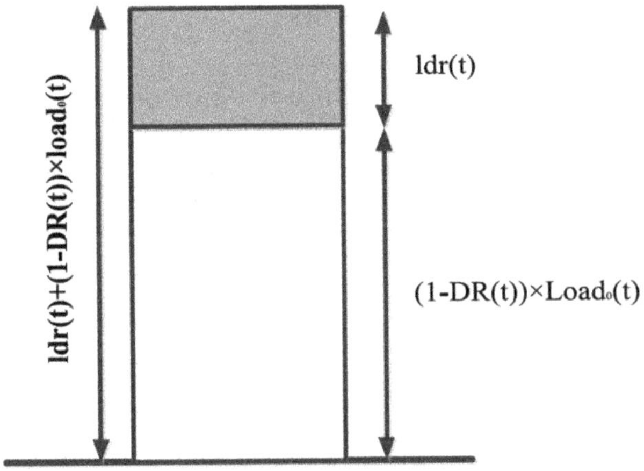

Fig. 2.3 Load modeling with TOU rate of DRP

$$\text{load}_t = (1 - \text{DR}_t) \cdot \text{load}_t^0 + \text{ldr}_t \tag{2.22}$$

$$\text{load}_t^0 - \text{load}_t = \text{DR}_t \cdot \text{load}_t^0 - \text{ldr}_t \tag{2.23}$$

where load_t is load after DRP performance in the same period.

Participation of the large consumer is presented by DR_t factor. Transferred load amount depends on the power price in the pool market. Related constraints to the DRP are expressed by constraints (2.24)–(2.27). At each time, the moveable load has variable size, which is presented by DR_t in constraint (2.24). The load increase in each time is limited using constraints (2.25). The maximum amount of DR_t and inc_t in each time is limited by constraints (2.26) and (2.27). In this chapter, DR_{max} and inc_{max} are considered 20%.

$$\sum_{t=1}^{T} \text{ldr}_t = \sum_{t=1}^{T} \text{DR}_t \cdot \text{load}_t^0 \tag{2.24}$$

$$\text{load}_t^{\text{inc}} \leq \text{inc}_t.\text{load}_t^0; \quad \forall t \tag{2.25}$$

$$\text{DR}_t \leq \text{DR}_{\text{max}}; \quad \forall t \tag{2.26}$$

$$\text{inc}_t \leq \text{inc}_{\text{max}}; \quad \forall t \tag{2.27}$$

where

inc_t is the size of load increase at time t
inc_{max} is the maximum size of load increase in each time t
DR_{max} is the maximum size of load participation in DRP

2.3 Simulation Results

In this section, the deterministic-based power procurement problem of a large consumer is solved. Proposed mixed-integer linear programming (MILP) formulation is solved by GAMS optimization [22] method using the CPLEX solver [23].

2.3.1 *Input Data*

Table 2.1 shows the daily demand levels of the large consumer, which is assumed as three load levels including valley, shoulder, and peak periods.

Information of 12 bilateral contracts containing the maximum and minimum quantity of energy with energy price and time levels are provided in Table 2.2. Among the 12 bilateral contracts, according to Table 2.2, six contracts are considered available for the all load levels, while the other six contracts are only assumed for peak load levels.

As said before, three pricewise linear models are considered to model SG units [24]. The information of SG units including the minimum and maximum power and ramping up and down times for different SG units is presented in Table 2.3. Table 2.4 presents the forecasted daily irradiation, temperature, and wind speed for a sample day which is required to calculate power output of WT and PV system.

Table 2.1 Classification of daily load levels

Level	Hours of the day
Valley (*V*)	1, 2, 3, 4, 5, 6, 7, 8
Shoulder (*S*)	9, 10, 15, 16, 17, 18, 23, 24
Peak (*P*)	11, 12, 13, 14, 19, 20, 21, 22

Table 2.2 Bilateral contract specification

Contract number	Validity level	Min. (MW)	Max. (MW)	Price ($/MWh)
1	*V, S, P*	0.5	2	75.84
2	*P*	0.8	1	109.16
3	*V, S, P*	0.6	2	119.55
4	*P*	0.5	1.5	82.76
5	*V, S, P*	0.8	2	109.24
6	*P*	0.6	1.5	113.60
7	*V, S, P*	0.5	2	88.12
8	*P*	0.2	1.5	83.52
9	*V, S, P*	0.4	2.5	110.64
10	*P*	0.6	2	94.76
11	*V, S, P*	0.5	2	86.8
12	*P*	0.8	2.5	111.84

Table 2.3 Self-generation units' data

Parameters	First unit	Second unit	Third unit	Units
Maximum power output	4	7	10	MW
Minimum power output	0	0	0	MW
S_1^{DG}	40	45	50	\$/MWh
S_2^{DG}	70	80	85	\$/MWh
S_3^{DG}	100	110	120	\$/MWh
P_1^{MAX}	2	4	7	MW
P_2^{MAX}	3	5	8	MW
P_3^{MAX}	4	7	10	MW
MUT_j	2	2	2	H
MDT_j	2	2	2	H
R_j^{up}	4	4	5	MW/h
R_j^{down}	4	4	5	MW/h

Table 2.4 Forecasted daily wind speed, temperature, and insulation for a sample day

Time (h)	Wind speed (m/s)	Temperature (°C)	Insulation (W/m^2)
1	10.5	24.7	0
2	13.5	24.5	0
3	14.9	24.3	0
4	15.6	24.4	0
5	19.5	24.5	93.5
6	20.6	26.5	219
7	14.4	27.5	467.5
8	14.1	28	637.5
9	11.3	28.5	780
10	9.7	28.8	916
11	7.0	29	1100
12	5.9	29.7	1033
13	8.9	29.8	850
14	9.5	30	680
15	10.4	29.8	595
16	8.8	29.5	255
17	7.1	29	212.5
18	8.3	27.7	153
19	9.9	26.5	63
20	7.5	24.8	0
21	8.8	25	0
22	9.8	24.8	0
23	9.2	24.6	0
24	8.4	24.8	0

Table 2.5 Information of WT/PV system parameter

Wind turbine			PV system		
Parameters	Values	Units	Parameters	Values	Units
p_r	2.05	MW	$P^M_{\text{Max},0}$	1.1	MW
V_{ci}	2	m/s	G_{a_0}	1000	W/m^2
V_r	14	m/s	$T_{M,0}$	25	°C
V_{c0}	25	m/s	NOCT	44	Years
C_t^{wind}	0	\$/MWh	C_t^{PV}	0	\$/MWh

Table 2.6 Information of ESS

Parameters	Values	Units
$P^{\max}_{\text{charge}}$	0.2	MW
$P^{\max}_{\text{disc}}$	0.15	MW
$X_b^{\max}$	2	MW
$X_b^{\min}$	0.2	MW
χ	90	%
η	80	%
C_t^B	0	\$/MWh

Finally, the required information of ESS and PV/WT are provided in Tables 2.5 and 2.6, respectively.

Figure 2.4 illustrates the forecasted demand of the large consumer for the study horizon. According to Fig. 2.4, the maximum demand occurs at hour 20, which is equal to 40.5. Furthermore, the profile of power price in the pool market is depicted in Fig. 2.5 [25]. The minimum power price is experienced at hour 13 with the price of 56 \$/MWh, while maximum power price is recorded at hour 20 which is equal to 116 \$/MWh.

2.3.2 *Deterministic Energy Procurement Results*

As discussed before, the power procurement problem is solved in deterministic case. Obtained result can be required by some uncertainty modeling approaches such as the two-point estimate method. In this case, the power price at the pool market is considered as fixed values. In order to demonstrate the impact of DRPs on the power procurement problem, two cases are considered as follows:

Case 1: The problem is solved without considering DRP.
Case 2: The problem is solved considering TOU rate of DRP.

Obtained results of expected cost of different power procurement sources including BCs, pool market, and operating cost of PV, WT, SG, and ESS systems are

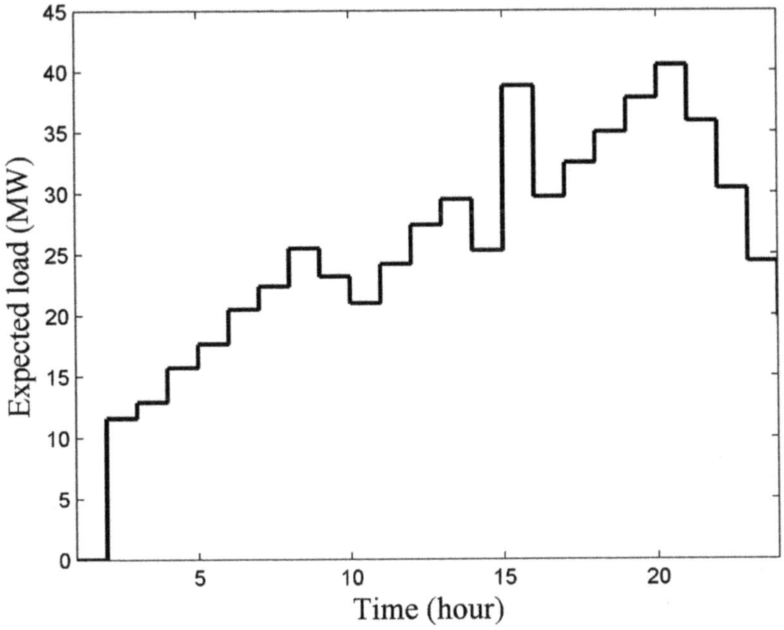

Fig. 2.4 Estimated load profile of the consumer

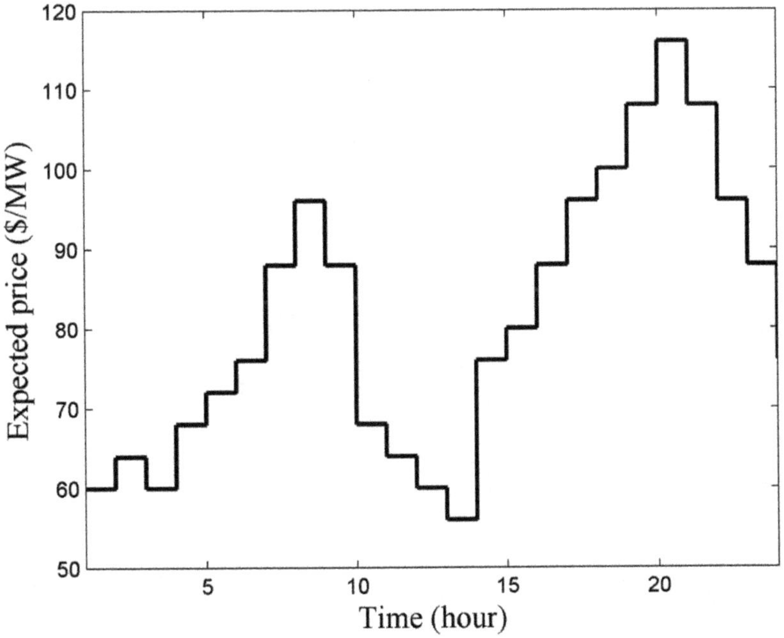

Fig. 2.5 Estimated market price

compared through two cases. Table 2.7 presents procured cost from different power procurement cost besides the total power procurement cost in two cases.

The total power procurement cost is equal to $40,253.252 in the first case which is solved without considering DRP. This amount is reduced to $38,401.038 after applying DRP. Comparing results of total procurement in two cases shows about 4.6% reduction in the procurement cost. Without considering DRP, the procured power cost from the pool market is equal to $13,168.363, which is reduced to $11,350.531 after implementing DRP. The cost of procured power from the BCs is $9322.080 in without and with considering DRP. Finally, procurement cost from SG units in the first case is $17,762.808, which is reduced to $34.38 in the second case.

Figure 2.6 shows the generated power from PV. The maximum output power of PV is 1.2 MW, which is recorded at hour 13. The minimum power generation of the PV is equal to 0 MW, which is recorded at hours 1–6 and 21–24. It is obvious that the PV system produces power when the solar radiation of the sun is available.

Table 2.7 Simulation results of deterministic approach in two cases

Different costs	Case 1	Case 2
Procurement cost from pool market ($)	13,168.363	11,350.531
Procurement cost from BCs ($)	9322.080	9322.080
Procurement cost from SG units ($)	17,762.808	17,728.427
Total procurement cost ($)	40,253.252	38,401.038

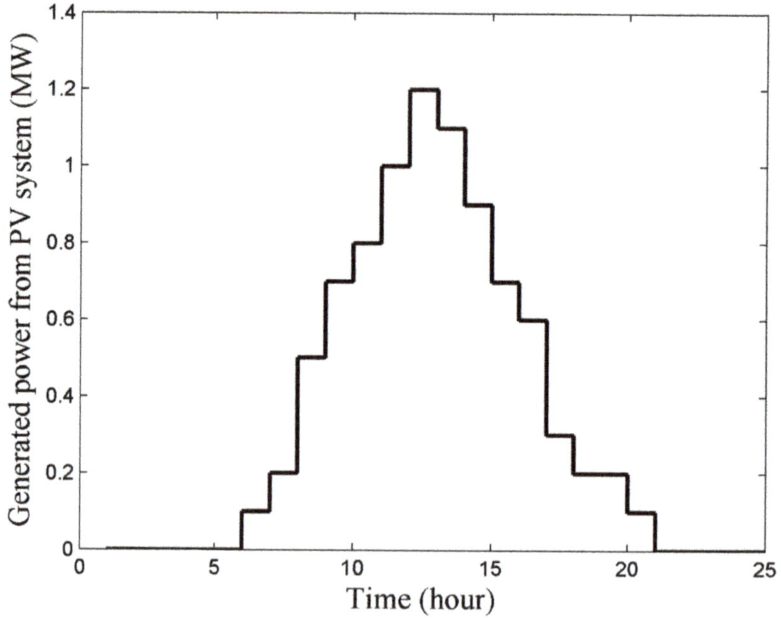

Fig. 2.6 Generated power by PV

Figure 2.7 presents the produced power by the WT. The maximum output power of WT is 2.1 MW, which is recorded at between hours 4 and 9. Furthermore, the minimum power generation of the WT is equal to 0.1 MW, which is recorded at hour 13.

Figure 2.8 illustrates the load profile in two cases for without and with considering DRP. As was expected, after applying DRP, the load is shifted from high-price periods to low-price periods, which has made the load profile more flattened in comparison with the first case. It should be denoted that for each time, the maximum participation factor of DRP is limited to 20% of the initial load.

Figure 2.9 presents the purchased power from the pool market. In comparison with case 1, procured power from the pool market is reduced during peak periods in case 2 because after applying DRP, the load amount is reduced in the peak period. In addition, purchased power during off-peak periods is increased in the second case due to load increment at these periods.

Procured power from self-generating is shown in Fig. 2.10. It is obvious that there is a slight difference between the obtained power of self-generating units in two cases, which comes from characteristics and parameters of these units.

Figure 2.11 presents the total procured power from the bilateral contracts for two cases in the horizon of the study. According to this Fig. 2.11, the amount of procured power for considering with and without DRP is same. The maximum total procured power from BCs is 7 MW, which is recorded between hours 20 and 24. Furthermore, 4 MW is the minimum procured power from the sum of 12 different bilateral contracts.

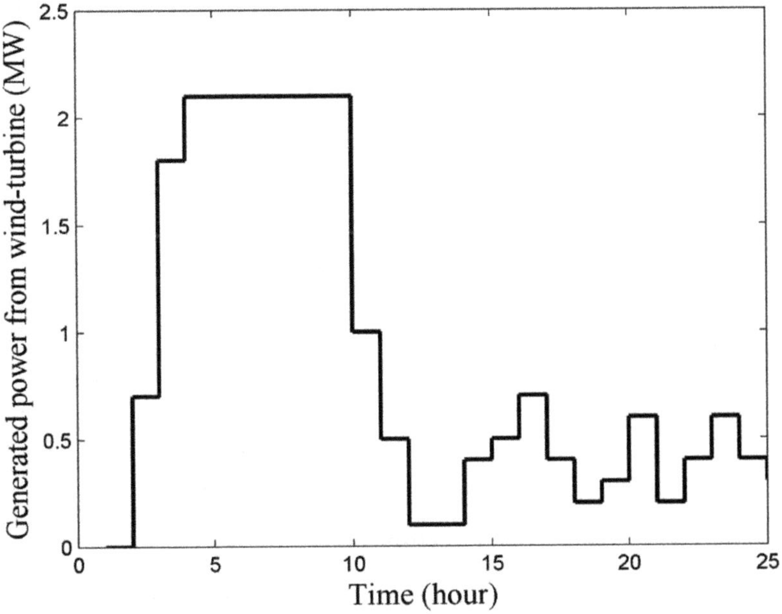

Fig. 2.7 Generated power by WT

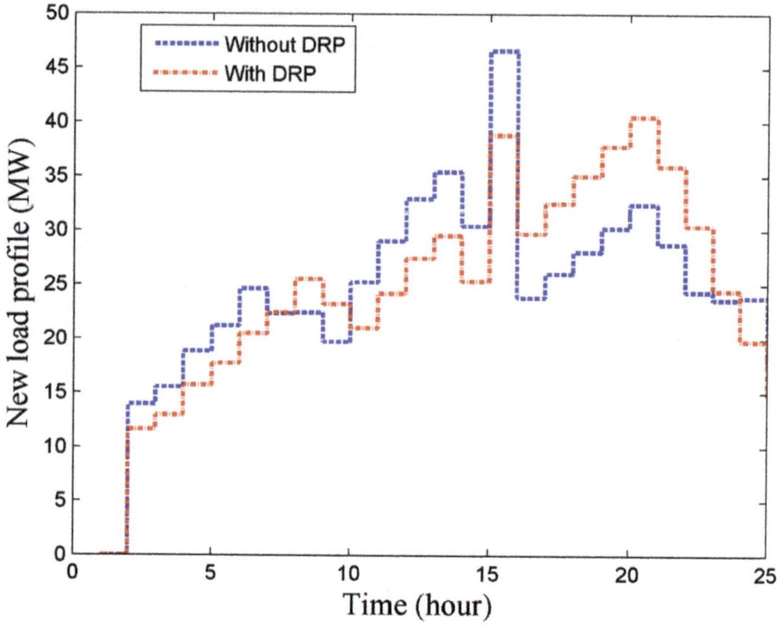

Fig. 2.8 Load profile considering with and without DRP

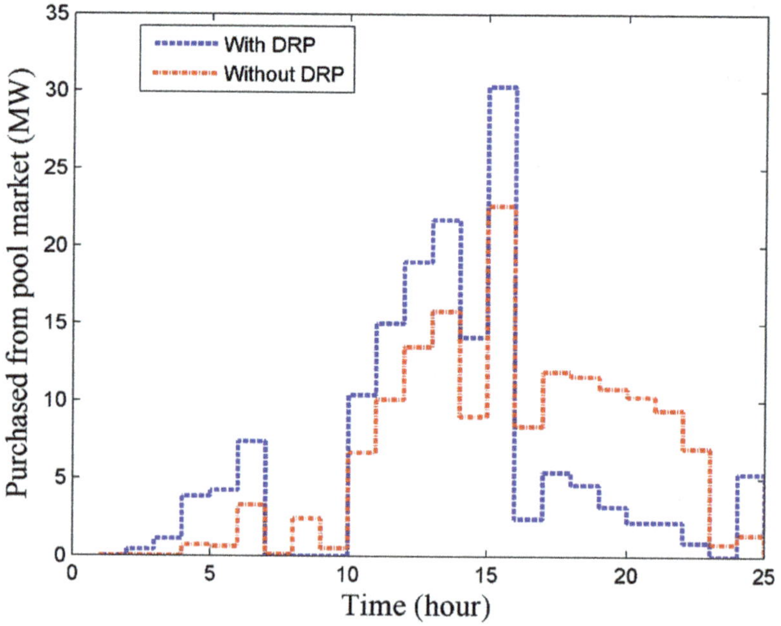

Fig. 2.9 Procured power from the pool market considering with and without DRP

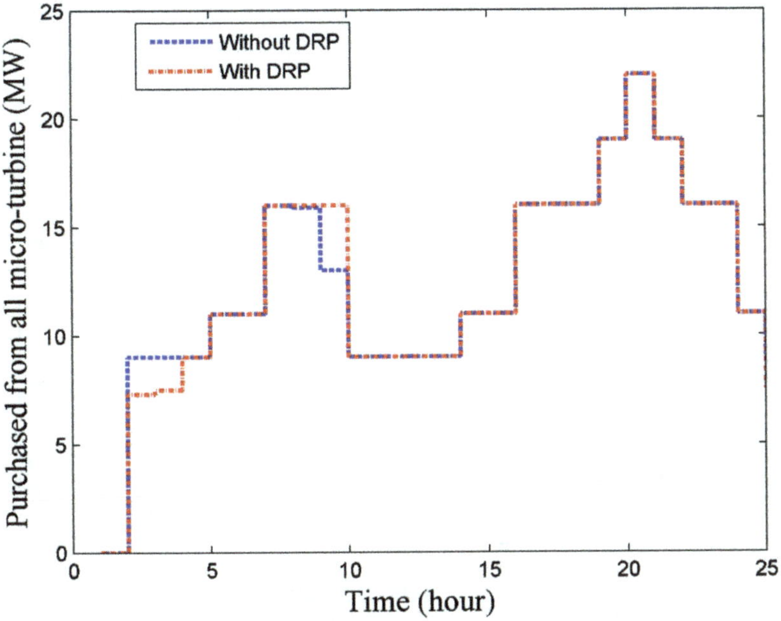

Fig. 2.10 Procured power from self-generating units considering with and without DRP

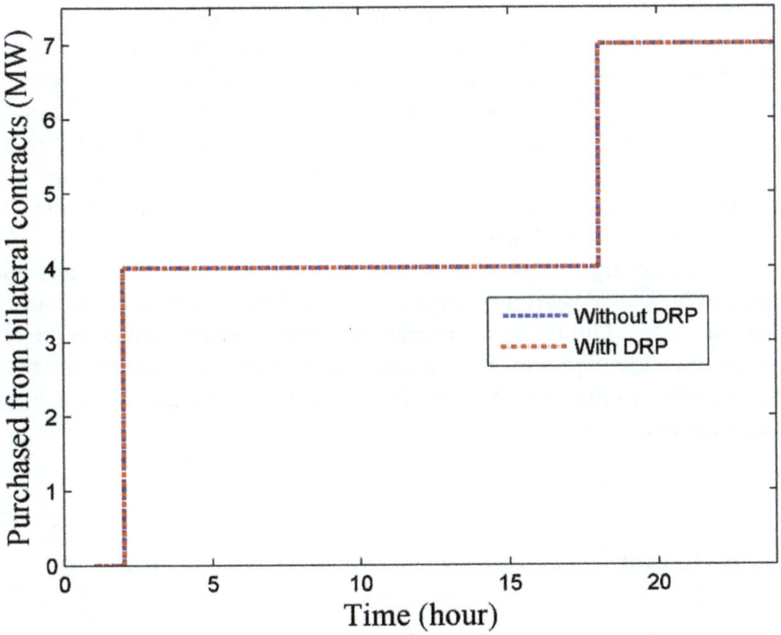

Fig. 2.11 Procured power from BCs considering with and without DRP

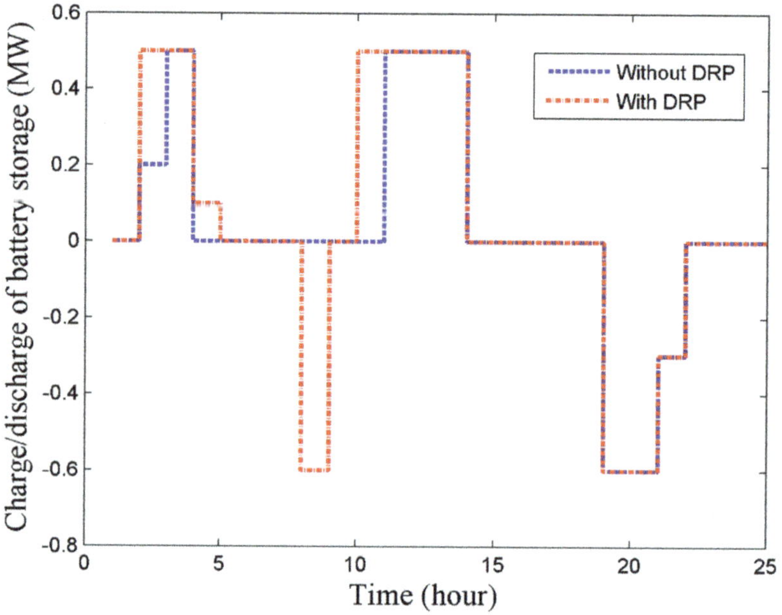

Fig. 2.12 Charging and discharging states of ESS considering with and without DRP

Participation of ESS in power procurement of large consumer for without and with considering DRP is depicted in Fig. 2.12. It should be denoted that charged power is presented by positive values, while discharged power is presented in negative values. The ESS is discharged at hour 8 in the second case because the DRP has increased load at this hour. In addition, the charging and discharging state of the ESS are the same between hours 11 and 24. The ESS is charged during hours 1–4 and 10–12, which are the low-price periods, and discharged during 19–22, which are the high-price and demand period.

Finally, stored energy in the ESS is illustrated for considering with and without DRP case in Fig. 2.13. As it can be seen, there is no difference between hours 10 and 24 in two cases. In addition, stored energy in between hours 1 and 9 in the second case, which considers applying demand response program, is more than the first case because according to Fig. 2.8, the DRP has transferred some part of load from this period to other periods.

2.4 Conclusion

In this chapter, the power procurement of a large consumer is solved considering the six alternative power sources including bilateral contracts, pool market, wind turbine, photovoltaic system, self-generating units, and energy storage systems in the

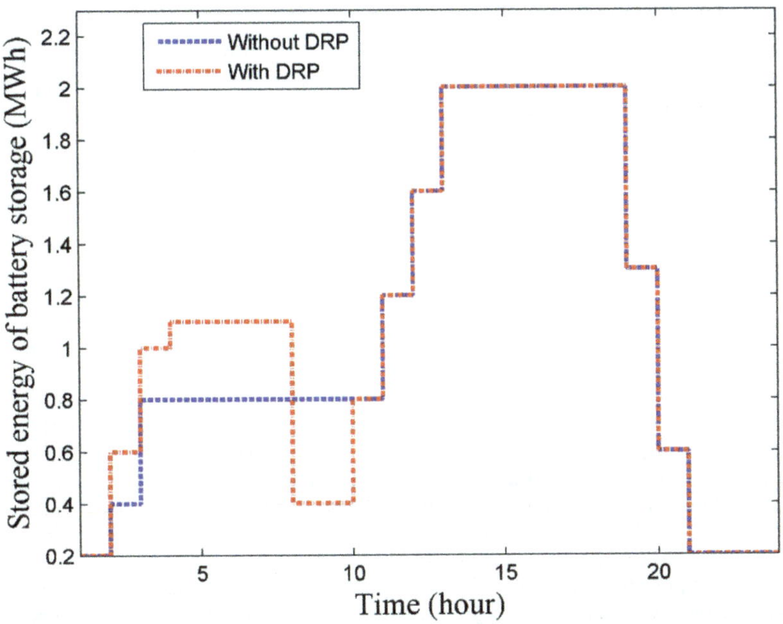

Fig. 2.13 Stored energy in the ESS considering with and without DRP

presence of the time-of-use rate of demand response program. The objective function and constraints are defined based on the abovementioned sources to minimize the total power procurement cost. Although, in the previous chapter, it has been said that power market has uncertainty in the pool market, the problem is solved using deterministic methods. To do so, power price in the pool market is considered as a predetermined value to get deterministic results. The time-of-use rate of demand response programs, as an effective way to reduce procurement cost, is implemented to decrease total power procurement cost. In order to investigate the impact of demand response program, the problem is solved in two discrete case studies as without and with considering demand response program. The problem is formulated as a mixed-integer linear program which is solved by the CPLEX solver under GAMS optimization software. Obtained results show that the total cost of power procurement is equal to \$40,253.252 and \$38,401.038 for without and with considering demand response program, respectively. These numbers show 4.6% reduction in power procurement cost because of implementing demand response program. Furthermore, obtained results show that procured power from the pool market is reduced to about 13.8%. In addition, operating cost of self-generating units obtained \$17,762.808 for without considering demand response program case. After implementing the demand response program, operating cost is reduced to \$17,728.427, which shows \$34.381 reduction. Finally, based on obtained results, the cost of procured power from bilateral contracts in the first case is equal to \$9322.080, which has not changed in the second case.

It should be denoted that the obtained results in this section can be used as input to some uncertainty modeling methods to investigate the uncertainty of power price in the market, load demand, or any other uncertainty in the problem. In Chap. 3, this problem is solved by modeling the uncertainty of power price and power output of renewable energy sources via stochastic programming method.

References

1. J.M. Morales, J. Perez-Ruiz, Point estimate schemes to solve the probabilistic power flow. IEEE Trans. Power Syst. **22**(4), 1594–1601 (2007)
2. P. Akbary, M. Ghiasi, M.R.R. Pourkheranjani, H. Alipour, N. Ghadimi, Extracting appropriate nodal marginal prices for all types of committed reserve. Comput. Econ., 1–26 (2017)
3. S. Nojavan, H. allah Aalami, Stochastic energy procurement of large electricity consumer considering photovoltaic, wind-turbine, micro-turbines, energy storage system in the presence of demand response program. Energy Convers. Manag. **103**, 1008–1018 (2015)
4. N. Sayyad, B. Mohammadi-Ivatloo, K. Zare, Optimal bidding strategy of electricity retailers using robust optimisation approach considering time-of-use rate demand response programs under market price uncertainties. IET Gener. Transm. Distrib. **9**(4), 328–338 (2015)
5. R. Rezaeipour, A. Zahedi, Multi-objective based economic operation and environmental performance of PV-based large industrial consumer. Sol. Energy **157**, 227–235 (2017)
6. H.A. Bagal, Y.N. Soltanabad, M. Dadjuo, K. Wakil, N. Ghadimi, Risk-assessment of photovoltaic-wind-battery-grid based large industrial consumer using information gap decision theory. Sol. Energy **169**, 343–352 (2018)
7. Y. Liu, W. Wang, N. Ghadimi, Electricity load forecasting by an improved forecast engine for building level consumers. Energy **139**, 18–30 (2017)
8. J.J. Vidal-Amaro, P.A. Østergaard, C. Sheinbaum-Pardo, Optimal energy mix for transitioning from fossil fuels to renewable energy sources – the case of the Mexican electricity system. Appl. Energy **150**, 80–96 (2015)
9. A.J. Chapman, B.C. McLellan, T. Tezuka, Prioritizing mitigation efforts considering co-benefits, equity and energy justice: fossil fuel to renewable energy transition pathways. Appl. Energy **219**, 187–198 (2018)
10. J. Benedek, T.-T. Sebestyén, B. Bartók, Evaluation of renewable energy sources in peripheral areas and renewable energy-based rural development. Renew. Sustain. Energy Rev. **90**, 516–535 (2018)
11. F. Tahri, A. Tahri, T. Oozeki, Performance evaluation of grid-connected photovoltaic systems based on two photovoltaic module technologies under tropical climate conditions. Energy Convers. Manag. **165**, 244–252 (2018)
12. H.A. Aalami, S. Nojavan, Energy storage system and demand response program effects on stochastic energy procurement of large consumers considering renewable generation. IET Gener. Transm. Distrib. **10**(1), 107–114 (2016)
13. W.-C. Wang, J.-J. Wang, W.T. Chong, The effects of unsteady wind on the performances of a newly developed cross-axis wind turbine: a wind tunnel study. Renew. Energy **131**, 644–659 (2018)
14. F. Mirzapour, M. Lakzaei, G. Varamini, M. Teimourian, N. Ghadimi, A new prediction model of battery and wind-solar output in hybrid power system. J. Ambient Intell. Humaniz. Comput., 1–11 (2017)
15. S. Nojavan, K. Zare, Interval optimization based performance of photovoltaic/wind/FC/electrolyzer/electric vehicles in energy price determination for customers by electricity retailer. Sol. Energy **171**, 580–592 (2018)

16. S. Nojavan, H. Ghesmati, K. Zare, Robust optimal offering strategy of large consumer using IGDT considering demand response programs. Electr. Power Syst. Res. **130**, 46–58 (2016)
17. M. Majidi, S. Nojavan, N. Nourani Esfetanaj, A. Najafi-Ghalelou, K. Zare, A multi-objective model for optimal operation of a battery/PV/fuel cell/grid hybrid energy system using weighted sum technique and fuzzy satisfying approach considering responsible load management. Sol. Energy **144**, 79–89 (2017)
18. D.T. Nguyen, L.B. Le, Optimal bidding strategy for microgrids considering renewable energy and building thermal dynamics. IEEE Trans. Smart Grid **5**(4), 1608–1620 (2014)
19. S. Nojavan, M. Majidi, N.N. Esfetanaj, An efficient cost-reliability optimization model for optimal siting and sizing of energy storage system in a microgrid in the presence of responsible load management. Energy **139**, 89–97 (2017)
20. S. Nojavan, K. Zare, B. Mohammadi-Ivatloo, Selling price determination by electricity retailer in the smart grid under demand side management in the presence of the electrolyser and fuel cell as hydrogen storage system. Int. J. Hydrogen Energy **42**(5), 3294–3308 (2017)
21. S. Nojavan, H. Qesmati, K. Zare, H. Seyyedi, Large consumer electricity acquisition considering time-of-use rates demand response programs. Arab. J. Sci. Eng. **39**(12), 8913–8923 (2014)
22. A. Brooke, D. Kendrick, A. Meeraus, GAMS user's guide. Redwood City (CA): The Scientific Press; 1990. Available: http://www.gams.com/docs/gams/GAMSUsersGuide.pdf
23. CPLEX 12 [Online]. Available: https://www.gams.com/latest/docs/S_CPLEX.html. Accessed 15 Jul 2018
24. K. Zare, M.P. Moghaddam, M.K. Sheikh El Eslami, Demand bidding construction for a large consumer through a hybrid IGDT-probability methodology. Energy **35**(7), 2999–3007 (2010)
25. H. Ebrahimian, S. Barmayoon, M. Mohammadi, N. Ghadimi, The price prediction for the energy market based on a new method. Econ. Res. Istraživanja **31**(1), 313–337 (2018)

Chapter 3
Stochastic-Based Energy Procurement

Mohammad Javad Golkar, Noradin Ghadimi, Sayyad Nojavan, and Mojtaba Madadkhani

Abstract In this section, the uncertainty of the pool price and load demand in the power procurement problem of a large consumer is modeled using the stochastic programming. Based on the deterministic formulation of the problem, which was introduced in Chap. 2, the stochastic formulation is presented to solve the problem.

Then, using normal distribution, uncertainties of the problem as the load, solar irradiance, temperature, and power price in the market are modeled generating 100 discrete scenarios considering the output power of photovoltaic panels. Furthermore, the uncertainty of the output power of the wind turbine is modeled using the Weibull distribution, which is used to model the wind speed by 100 scenarios.

Two different cases as without and with considering demand response programs are assumed to solve the problem and investigate the impact of demand response programs. To do so, the time-of-use rate of demand response is applied to reduce the total power procurement cost of the large consumer.

The problem is formulated as a mixed-integer linear program, which is solved by the CPLEX solver under GAMS optimization software.

M. J. Golkar
Department of Electrical and Computer Engineering, Zahedan Branch, Islamic Azad University, Zahedan, Iran

N. Ghadimi (✉)
Young Researchers and Elite Club, Ardabil Branch, Islamic Azad University, Ardabil, Iran
e-mail: noradin.ghadimi@iauardabil.ac.ir

S. Nojavan
Department of Electrical Engineering, University of Bonab, Bonab, Iran
e-mail: sayyad.nojavan@bonabu.ac.ir

M. Madadkhani
Department of Information and Communications Technology, Malek Ashtar University of Technology, Tehran, Iran

S. Nojavan et al. (eds.), *Robust Energy Procurement of Large Electricity Consumers*,
https://doi.org/10.1007/978-3-030-03229-6_3

3.1 Introduction

Uncertainty is an inevitable part of the system, which is under study to be optimized. Deterministic mode results can easily be nullified through deviation of nominal values of different parameters under various uncertain parameters as load demand and power price [1]. It is likely to happen even because of other unpredicted events such as natural disasters, malfunctioning of the system, or insufficient information about the considered system. To do so, uncertainty modeling in the decision-making process should be well studied. So, different approaches are introduced to model uncertainty in the understudy system.

In order to select the most appropriate methodologies among different developed methods, which are proposed for different purposes, enough knowledge about the characteristics of the system and method is required. As an example, multi-parametric programming [2] can be used to model uncertainty when there is a relatively small number of uncertain parameter. By using this method, a closed-form solution of the optimization problems can be obtained through mapping the parameter space. Also, this method can provide a robust estimation of the given solutions. As another example, if comprehensive probabilistic information about different parameters of the system is available, the stochastic programming method [3] can be considered as a solution to model uncertainty and optimize the performance of the system.

In this chapter, the power procurement problem of a large consumer is solved using the stochastic programming method to model the uncertainties of the power price in the pool market, load demand, and generation of renewable energy sources. In the stochastic programming method, the considered uncertainty is modeled through a set of discrete scenarios. The power price uncertainty in the pool market, which is the main power procurement source for a large consumer, is modeled using 100 discrete scenarios. In addition, a set of scenarios is generated for wind speed using the Weibull distribution to model the output power of the wind turbine. In addition, it should be noted that the output power of the photovoltaic panel is model generating 100 scenarios for solar irradiance and temperature using normal distribution.

The stochastic formulation of the power procurement problem is presented based on the deterministic case for a case study including 12 bilateral contracts with different predetermined prices for different periods, self-generating units, photovoltaic (PV) system, wind turbine (WT), and power market as sources to get the required demand of the large consumer. In addition, as discussed in Chap. 2, in order to investigate the impact of demand response programs (DRPs), the problem is solved in two different cases as without and with considering the time-of-use rate of DRP.

3.2 Stochastic Programming Method

In this section, the basic features of the stochastic programming method are introduced. Unlike the deterministic methods, which solve the problem considering known parameters, the stochastic programming method is used to formulate the optimization problems involving uncertain parameter to determine the impact of an unknown parameter at the time that a decision should be made [4, 5]. By considering some given set of possible values for the uncertain parameter, it is possible to seek for a feasible solution for all possible parameter choices and optimize a given objective function [6]. As an example, designing a bridge with the least weight using steel having a tensile strength can be considered. It should be noted that in the presented example, the tensile strength is considered known only within some tolerance. The stochastic programming method can be used in a setting that a decision should be made as one-off [7].

The two-stage stochastic programming model is one of the widely applied models [8], in which some decisions should be made at the first stage. In the second stage, a recourse decision is made by considering any undesired impact that might be resulted as an output of the first-stage decision [9]. For more information about the stochastic programming method, please refer to [10].

3.3 Case Study

In order to investigate the impact of uncertainties of load, power price, and power output of renewable energy sources in the power procurement problem, the stochastic programming method is applied on the case study, which is proposed in Chap. 2. It should be denoted that the normal distribution is used to generate 100 scenarios for load, power price, solar irradiance, and temperature. Also, wind speed scenarios are generated using the Weibull distribution. Using the wind speed, solar irradiance, and temperature scenarios, power output of WT and PV units is modeled.

The considered case study is presented in Fig. 3.1. According to Fig. 3.1, the large consumer has six power procurement sources. The problem is solved through various scenarios, and as it can be seen, the decision-maker has to select the optimal strategy among the provided solution based on his/her experience.

3.4 Stochastic Formulation of the Problem

In this section, the stochastic formulation of the problem is presented based on the deterministic formulation, which is presented in Chap. 2. For each scenario, the presented objective function is concentrated on minimizing the cost of procured power from different resources including pool market (PM), bilateral contracts

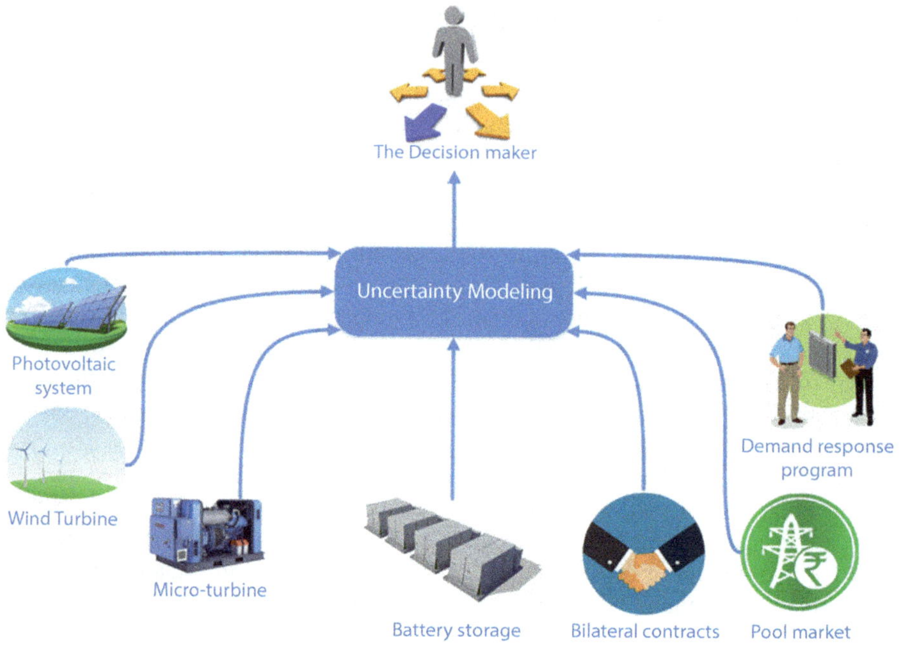

Fig. 3.1 Proposed case study of the power procurement problem

(BCs), photovoltaic system (PV), wind turbine (WT), self-generation units, and electrical storage systems (ESSs). It should be denoted that proposed formulation has applied 100 discrete scenarios. The decision-maker has to make optimal decisions among obtained results of the uncertainty environment. To do so, there are two kinds of decisions, which are distinguished in the decision-making process [11]:

1. *The first-stage or here-and-now decisions*: the decisions are made prior to the realization of the stochastic process. Therefore, the variables, which represent the here-and-now decisions, do have any dependency on a realization of the stochastic process. In this particular case, related variables to the BCs are classified as first-stage or here-and-now.
2. *The second-stage or wait-and-see decisions*: this refers to the decisions which are made after knowing the actual realization of the stochastic process. Hence, these decisions depend on each realization vector of the stochastic process. In the situations that a set of discrete scenarios presents the stochastic process, for each realization of the scenarios, a second-stage decision variable is defined. In the power procurement problem of the large consumer, the decision variables related to procured power from PV, WT, ESS, self-generating units, pool market, and DRP are the second-stage variables.

3.4.1 Objective Function

Based on Eq. (2.1) which presents the objective function of power procurement problem of a large consumer in the deterministic mode, Eq. (3.1) models the objective function in the stochastic case.

$$\underset{P^{p}_{t,s},P^{\text{BC}}_{l,t},P_{j,h,t,s},P^{\text{charge}}_{t,s},P^{\text{disc}}_{t,s},P^{\text{wind}}_{t,s},P^{\text{PV}}_{t,s}}{\text{Minimize}} \quad \sum_{l=1}^{B}\sum_{t=1}^{T}\lambda_{l,t}P^{\text{BC}}_{l,t}$$

$$+\sum_{t=1}^{T}\sum_{s=1}^{N_s}P_s\times\left\{\begin{array}{l}\lambda_{t,s}P^{p}_{t,s}+\sum_{j=1}^{N_j}\sum_{h=1}^{N_h}S^{\text{DG}}_{j,h}P_{j,h,t,s}\\+C^{B}_{t}\times\left(\chi\times P^{\text{charge}}_{t,s}+\dfrac{P^{\text{disc}}_{t,s}}{\eta}\right)\\+C^{\text{wind}}_{t}\times P^{\text{wind}}_{t,s}+C^{\text{PV}}_{t}\times P^{\text{PV}}_{t,s}\end{array}\right\} \tag{3.1}$$

where

s is the index of scenario
N_s is the number of scenarios
P_s is the probability of each scenario s
$\lambda_{t,s}$ is the market price at time t in scenario s [\$/MWh]
$P^{p}_{t,s}$ is the purchased power from the power market at time t in scenario s [MW]
$P_{j,t,s}$ is the produced power from the jth unit of the self-generation units at time t in scenario s [MW]
$P^{\text{charge}}_{t,s}, P^{\text{disc}}_{t,s}$ is the charging/discharging power of battery storage (MW) at time t in scenario s [MW]
$P^{\text{wind}}_{t,s}$ is the produced power from the wind turbine at time t in scenario s [MW]
$P^{\text{PV}}_{t,s}$ is the produced power from the PV system at time t in scenario s [MW]

3.4.2 Power Balance Constraint

Equation (3.2) presents the load balance between the load demand and procured power from alternative power resources, which is obtained based on Eq. (2.2). In order to predict the load demand, the proposed method in [12] can be considered.

$$\sum_{l=1}^{B} P_{l,t}^{\mathrm{BC}} + P_{t,s}^{p} + \sum_{j=1}^{N_j}\sum_{h=1}^{N_h} P_{j,h,t,s} + P_{t,s}^{\mathrm{wind}} + P_{t,s}^{\mathrm{PV}} + P_{t,s}^{\mathrm{disc}} = \left(\left(1 - \mathrm{DR}_{t,s}\right)\cdot \mathrm{load}_{t,s}^{0} + \mathrm{ldr}_{t,s}\right) + P_{t,s}^{\mathrm{charge}}; \quad \forall t,s \tag{3.2}$$

where

$\mathrm{DR}_{t,s}$ is the potential of DRP implementation at time t in scenario s
$\mathrm{load}_{t,s}^{0}$ is the initial load at time t in scenario s
$\mathrm{ldr}_{t,s}$ is the shifted load by DRP at time t in scenario s

3.4.3 Constraints of Bilateral Contracts

Procured power from BCs should be in an acceptable interval which is described by Eq. (3.3). Total procured power from different BCs is summed in Eq. (3.4).

$$P_{l,t}^{\min} s_l < P_{l,t}^{\mathrm{BC}} < P_{l,t}^{\max} s_l; \quad \forall l,t \tag{3.3}$$

$$P_{t}^{\mathrm{BC}} = \sum_{l=1}^{N_l} P_{l,t}^{\mathrm{BC}}; \quad \forall t \tag{3.4}$$

where

$P_{l,t}^{\min}$ is the minimum capacity relating to contracts l at time t
$P_{l,t}^{\max}$ is the maximum capacity relating to contracts l at time t
s_l is the binary variable, "1" if bilateral agreement l is selected and 0 otherwise
P_{t}^{BC} is the total procured power from the bilateral contracts at time t

3.4.4 Model of Self-Generation Units

Three piecewise linear curves are used to model the operating cost of SG units [13], which is shown in Fig. 2.1. The operating cost of SG units for each scenario is modeled by Eq. (3.5). Equations (3.6)–(3.11) present the constraints of SG units in which the sum of procured energy is limited by constraints (3.6) and (3.7); the ramping up and down rate limits are expressed by constraints (3.8) and (3.9), respectively; and the minimum up and down time constraints are presented by the auxiliary variables $\mathrm{Up}_{i,j}$ and $\mathrm{Dn}_{i,j}$ which are defined by constraint (3.11). It should be noted that constraints (3.6)–(3.11) are based on the constraints (2.6)–(2.11) which are used in Chap. 2.

$$\text{Cost} = \sum_{t=1}^{T} \sum_{j=1}^{N_j} \sum_{h=1}^{N_h} \sum_{s=1}^{N_s} P_s \times S_{j,h}^{\text{DG}} P_{j,h,t,s} \tag{3.5}$$

$$0 \leq P_{j,h,t,s} \leq \left(P_{j,h}^{\text{MAX}} - P_{j,h-1}^{\text{MAX}}\right) \times U_{j,t}; \quad \forall j, h, t, s \tag{3.6}$$

$$0 \leq P_{j,1,t,s} \leq P_{j,1}^{\text{MAX}} \times U_{j,t}; \quad \forall j, t, s \tag{3.7}$$

$$\sum_{h=1}^{N_h} P_{j,h,t} - \sum_{h=1}^{N_h} P_{j,h,t-1} \leq R_j^{\text{up}} \times U_{j,t}; \quad \forall j, t \tag{3.8}$$

$$\sum_{h=1}^{N_h} P_{j,h,t-1} - \sum_{h=1}^{N_h} P_{j,h,t} \leq R_j^{\text{down}} \times U_{j,t-1}; \quad \forall j, t \tag{3.9}$$

$$\begin{aligned} U_{j,t} - U_{j,t-1} &\leq U_{j,t+\text{Up}_{j,i}}; \quad \forall j, \forall t, \forall i \\ U_{j,t-1} - U_{j,t} &\leq 1 - U_{j,t+\text{Dn}_{j,i}}; \quad \forall j, \forall t, \forall i \end{aligned} \tag{3.10}$$

$$\begin{aligned} \text{Up}_{j,i} &= \begin{Bmatrix} i & i \leq \text{MUT}_j \\ 0 & i > \text{MUT}_j \end{Bmatrix} \\ \text{Dn}_{j,i} &= \begin{Bmatrix} i & i \leq \text{MDT}_j \\ 0 & i > \text{MDT}_j \end{Bmatrix} \end{aligned} \tag{3.11}$$

3.4.5 *Wind Turbine Model*

Equation (3.12) presents the maximum output power produced by WT in each time t and scenario s, which is obtained based on constraint (3.12). As said before, the uncertainty of the wind speed can be modeled using the forecast error curves of wind speed which are obtained based on the Weibull distribution [14, 15].

$$P_{t,s}^{\text{wind, max}} = \begin{cases} 0 & V_{t,s}^{w} < V_{\text{ci}} \\ p_r \times \left(\frac{V_{t,s}^{w} - V_{\text{ci}}}{V_r - V_{\text{ci}}}\right)^3 & V_{\text{ci}} < V_{t,s}^{w} < V_{\text{cr}} \\ p_r & V_r < V_{t,s}^{w} < V_{c0} \\ 0 & V_{t,s}^{w} > V_{c0} \end{cases} \tag{3.12}$$

where

$P_{t,s}^{\text{wind, max}}$ is the maximum available power wind turbine at time t in scenario s
$V_{t,s}^{w}$ is the wind speed at time t

Procured energy from WT at each time t and scenario s is limited by using Eq. (3.13).

$$P_{t,s}^{\text{wind}} \leq P_{t,s}^{\text{wind, max}}; \quad \forall t, s \tag{3.13}$$

3.4.6 Photovoltaic System Model

For each scenario s and time t, the maximum procured power from PV systems is presented in Eq. (3.14) [16, 17].

$$P_{t,s}^{M,\max} = \frac{G_{t,s}^{a}}{G_{a_0}} \times \left\{ P_{\text{Max},0}^{M} + \mu_{P\max} \times \left(T_{t,s}^{a} + G_{t,s}^{a} \times \frac{\text{NOCT} - 20}{800} - T_{M,0} \right) \right\} \tag{3.14}$$

where

$P_{t,s}^{M,\max}$ is the maximum power at any temperature T_t^a for time t and scenario s
$G_{t,s}^{a}$ is the insulation for each time t and scenario s

Procured power from PV systems is limited using Eq. (3.15).

$$P_{t,s}^{\text{PV}} \leq P_{t,s}^{M,\max}; \quad \forall t, s \tag{3.15}$$

3.4.7 Energy Storage System Model

Equations (3.16)–(3.21) present the modeling of the energy storage system (ESS), which is used as a power source for the large consumer. It should be denoted that these equations are obtained based on deterministic formulations in Eqs. (2.16)–(2.21). Equation (3.16) presents the initial energy of the ESS. In addition, for each scenario s and time t, the charging and discharging power are limited by constraints (3.17) and (3.18), respectively; the level of stored energy in the ESS is presented by Eq. (3.19). Furthermore, simultaneously charging and discharging of the ESS are prevented by constraint (3.20); the energy dynamic model is described by constraint (3.21), for each scenario s and time t.

$$X_{t_0}^{b} = X_0^{b} \tag{3.16}$$

$$P_{t,s}^{\text{charge}} \leq P_{\text{charge}}^{\max} \times U_{t,s}^{\text{charge}}; \quad \forall t, s \tag{3.17}$$

$$P_{t,s}^{\text{disc}} \leq P_{\text{disc}}^{\max} \times U_{t,s}^{\text{disc}}; \quad \forall t, s \tag{3.18}$$

$$X_b^{\min} \leq X_{t,s}^b \leq X_b^{\max}; \quad \forall t, s \tag{3.19}$$

$$U_{t,s}^{\text{charge}} + U_{t,s}^{\text{disc}} \leq 1; \quad \forall t, s \tag{3.20}$$

$$X_{t,s}^b = X_{t-1,s}^b + \chi \times P_{t,s}^{\text{charge}} - \frac{P_{t,s}^{\text{disc}}}{\eta}; \quad \forall t, s \tag{3.21}$$

where

$P_{t,s}^{\text{charge}}/P_{t,s}^{\text{disc}}$ is the charging/discharging power of the ESS (MW) at time t and scenario s

$U_{t,s}^{\text{charge}}/U_{t,s}^{\text{disc}}$ is the binary variable, which is equal to one if charging/discharging of the ESS at time t and scenario s

$X_{t,s}^b$ is stored energy in the ESS at time t in scenario s

3.4.8 Demand Response Program Model

The time-of-use (TOU) rate of DRP is considered to reduce the power procurement cost of the large consumer. The TOU rate of DRP has a limitation in the maximum shifting capability of the load, which is assumed to be 20% in this chapter. Equations (3.22) and (3.23) model the TOU-DRP for each scenario s and time t.

$$\text{load}_{t,s} = (1 - \text{DR}_{t,s}) \cdot \text{load}_{t,s}^0 + \text{ldr}_{t,s} \tag{3.22}$$

$$\text{load}_{t,s}^0 - \text{load}_{t,s} = \text{DR}_{t,s} \cdot \text{load}_{t,s}^0 - \text{ldr}_{t,s} \tag{3.23}$$

where

$\text{load}_{t,s}$ is the load after DRP performance in the same period for each scenario.

As said in Chap. 2, in TOU rate of DRP, the transferred load amount depends on the power price in the pool market. This is shown in constraints (3.24)–(3.27) for each scenario s and time t. These constraints are obtained based on constraints (2.24)–(2.27).

$$\sum_{t=1}^{T} \text{ldr}_{t,s} = \sum_{t=1}^{T} \text{DR}_{t,s} \cdot \text{load}_{t,s}^0; \quad \forall s \tag{3.24}$$

$$\text{load}_{t,s}^{\text{inc}} \leq \text{inc}_{t,s} \cdot \text{load}_{t,s}^0; \quad \forall t, s \tag{3.25}$$

$$\text{DR}_{t,s} \leq \text{DRmax}; \quad \forall t, s \tag{3.26}$$

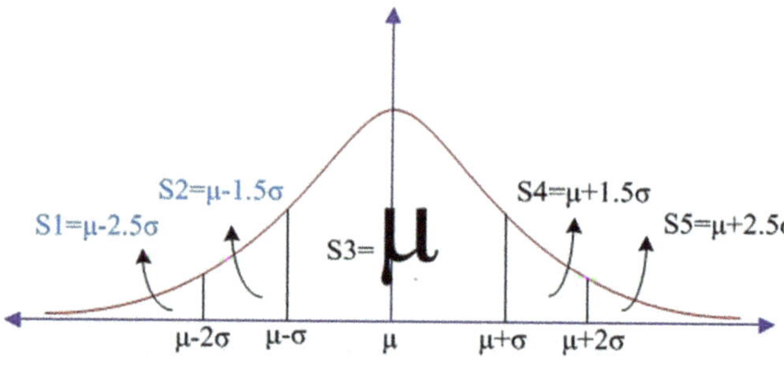

Fig. 3.2 Probability distribution function for uncertainty parameters

$$\text{inc}_{t,s} \leq \text{inc}_{\max}; \quad \forall t, s \tag{3.27}$$

where

$\text{inc}_{t,s}$ is the size of load increase at time t and scenario s
$\text{load}_{t,s}^{\text{inc}}$ is the load increase of each time t in scenario s

3.4.9 Uncertainty Modeling of Pool Price and Load Demand

The forecast error distribution curves, which are divided into some intervals with the width of one standard deviation, are used to model the uncertainty of the pool price and load demand [18]. The deterministic solution, which is discussed in Chap. 2, is used as an input of the uncertain model. The pool price and load demand variances are considered 10% of the average. Figure 3.2 presents a sample discrete form of the forecast error probability distribution function. Then, the probability of each scenario and average value of each interval should be calculated [19]. Note that the probability of each scenario is obtained by integration of the area under the probability distribution curve in each interval, and the average value of each interval is considered as the realized forecast error in the corresponding scenario.

3.5 Numerical Simulation

In this section, the power procurement problem faced by a large consumer is solved using a stochastic programming formulation. As considered in Chap. 2, the problem is formulated as a mixed-integer linear program (MILP) which is solved by the CPLEX solver [20] under GAMS optimization software [21].

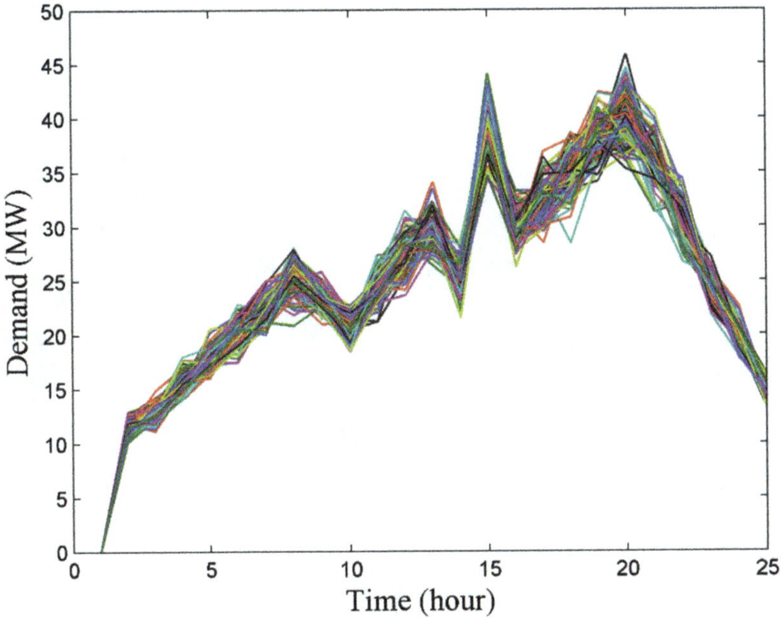

Fig. 3.3 The load profile of the consumer considering 100 scenarios

3.5.1 *Input Data*

The daily load demand of the large consumer is divided into three load levels as valley, shoulder, and peak period, which is detailed in Table 2.1.

Twelve bilateral contracts with predetermined prices and periods are considered to supply the required demand by the large consumer. The required information of these contracts is presented in Table 2.2. In addition, Table 2.3 presents the data of the self-generating units. Forecasted daily wind speed, temperature, and insulation for a sample day are the required information of ESS and PV/WT, which are provided in Tables 2.4–2.6, respectively.

In order to model the uncertainty of load demand, 100 scenarios are used to the forecasted demand of the large consumer, which is shown in Fig. 3.3. As it can be seen from this figure, in most of the scenarios, the maximum demand occurs at hour 20. In the same manner, the uncertainty of the pool market price is modeled using a set of 100 scenarios, which is depicted in Fig. 3.4.

3.5.2 *Stochastic Energy Procurement Results*

The power procurement problem is solved considering the uncertainty of the pool price and load demand using stochastic programming. The problem is solved

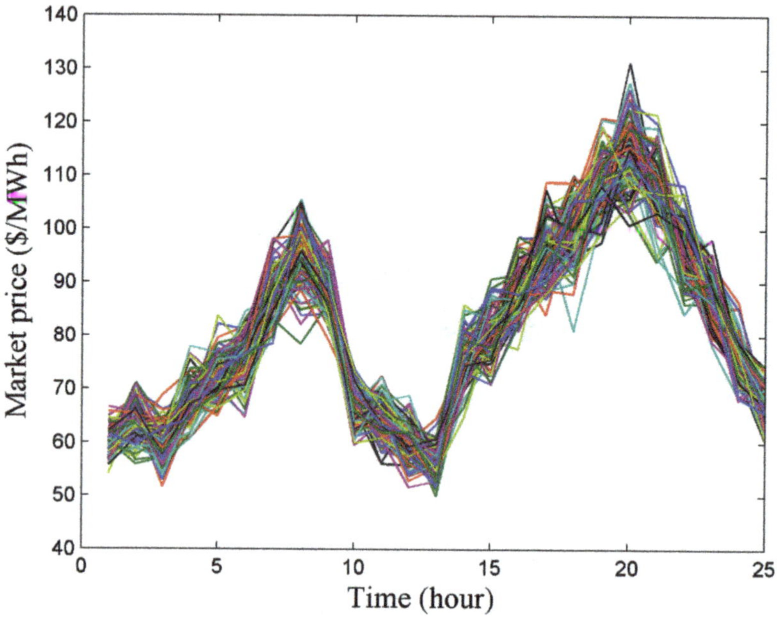

Fig. 3.4 Estimated market price scenarios

Table 3.1 Simulation results of deterministic approach in two cases

Different costs	Case 1	Case 2
Procurement cost from pool market ($)	12,827.042	11,241.042
Procurement cost from BCs ($)	9322.1	9322.080
Procurement cost from SG units ($)	18,237.456	17,964.255
Total procurement cost ($)	40,386.577	38,527.377

considering the two following cases to show the impact of TOU-DRP, as done in Chap. 2:

Case 1: The problem is solved without considering DRP
Case 2: The problem is solved considering TOU rate of DRP

It should be denoted that 100 scenarios are considered to model the uncertainty in the system. The results are presented for scenario 50. Table 3.1 provides the procured cost of different power procurement sources.

According to Table 3.1, the cost of procured power from the pool market is $12,827.042, which decreased to $11,241.042 by applying the demand response program, which shows a 12.36% reduction in procured power cost. The cost of purchased power from the bilateral contract is obtained as $9322.080 for both cases. Due to predetermined prices of the bilateral contracts, the participation of these contracts has not changed. The operating cost of self-generating units which is considered as the power procurement cost is obtained as $18,237.456 and

$17,964.255 for without and with considering demand response programs, respectively. Finally, the total power procurement cost is obtained as $40,386.577 without considering the demand response program. By applying the time-of-use rate of demand response program, the total power procurement cost is equal to $38,527.377. These numbers show a 4.6% reduction in power procurement cost.

Figure 3.5 presents the output power of PV system for different considered scenarios, which are used to model the uncertainty of the system. The maximum output power of PV is 1.36 MW, which is recorded at hour 12 in scenario 60. In addition, the minimum power output of the PV is equal to 0 MW recording at hours 1–5 and 21–24 in all scenarios.

Generated power by the WT is illustrated in Fig. 3.6. The maximum output power of WT is 2.58 MW, which is recorded at hour 5 in scenario 27. Furthermore, the minimum power generation of the WT is equal to 0.05 MW, which is recorded at hours 13 and 18 in scenarios 16, 25, 78, 85, and 93.

The obtained load profile in scenario 50 is shown in Fig. 3.7 for without and with considering DRP. The demand response program has transferred some part of the load during peak periods which are experienced between 17 and 22 to other periods as 2–7. It is obvious that DRP has flattened the load curve resulting to lower procurement cost. Note that the maximum participation factor of DRP is considered as 20% of the base load.

Purchased power from the pool market in scenario 50 is shown in Fig. 3.8 for two considered cases. In the first case, the total purchased power from the pool market is

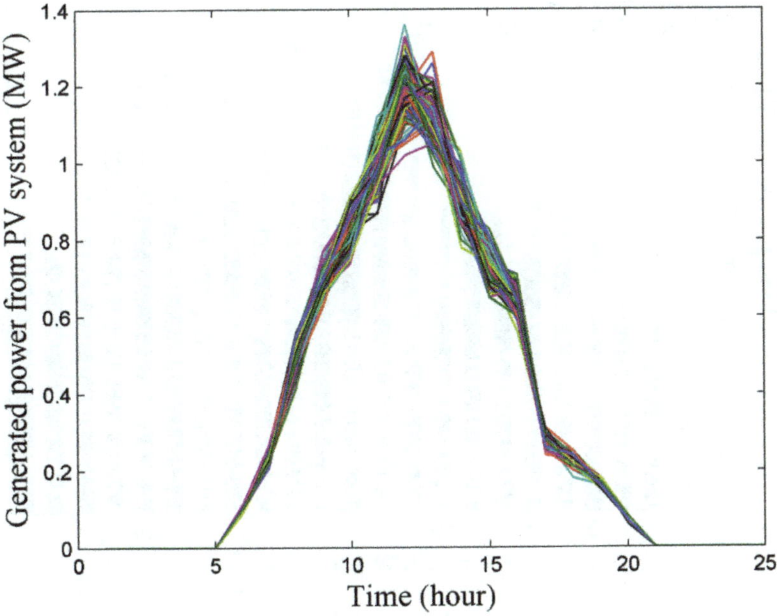

Fig. 3.5 Generated power by PV systems for different scenarios

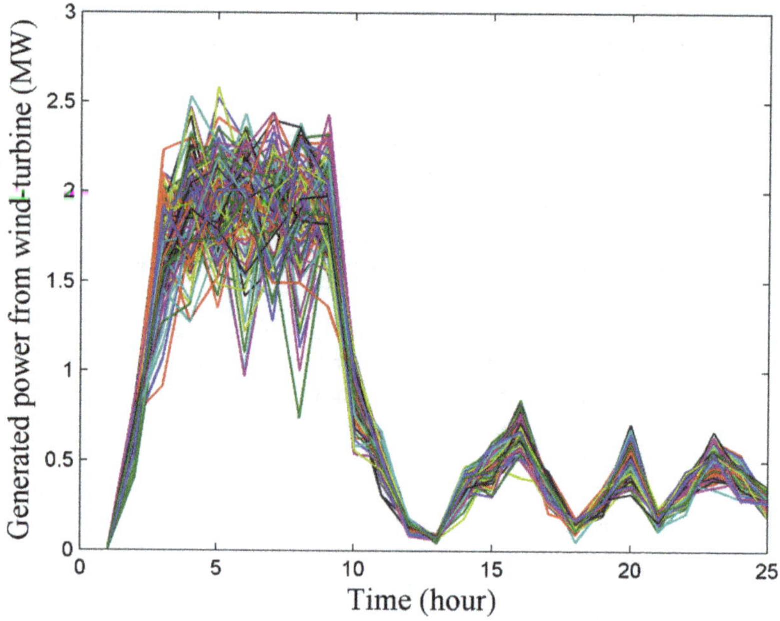

Fig. 3.6 Generated power by WT for different scenarios

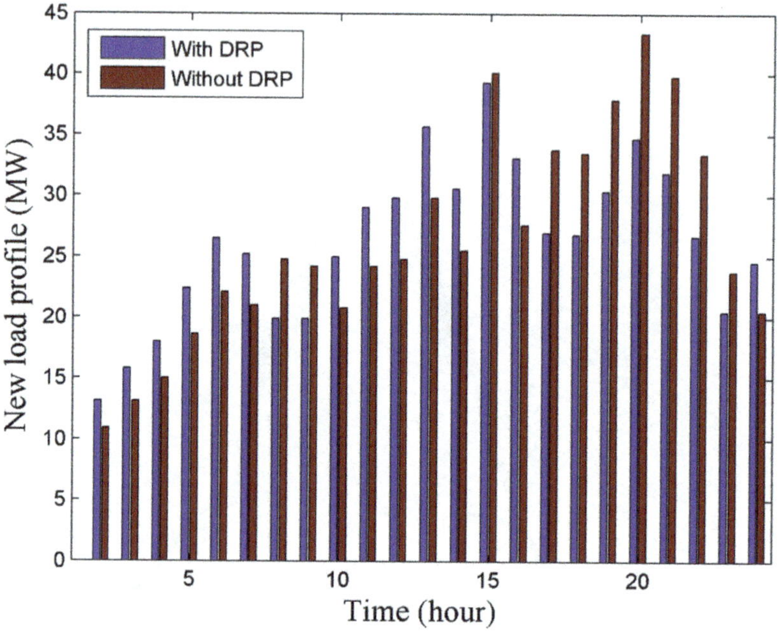

Fig. 3.7 Load profile considering with and without DRP for scenario 50

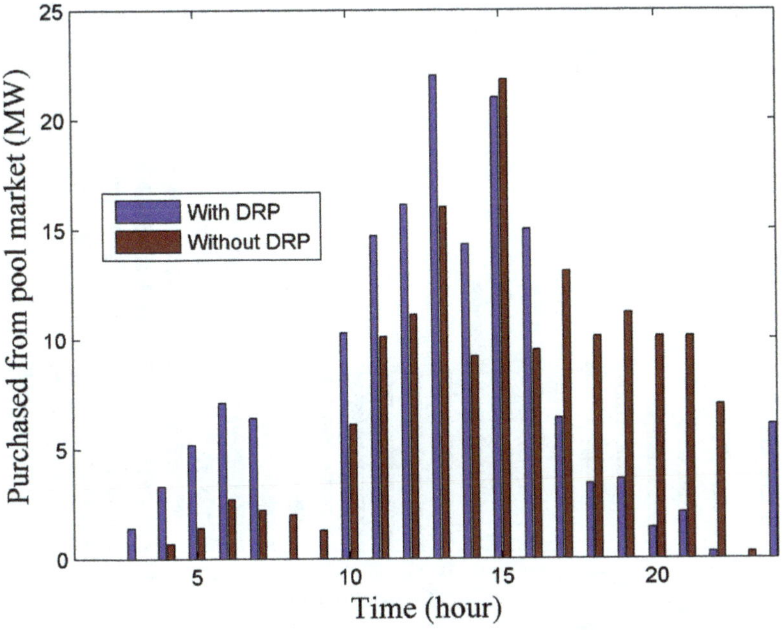

Fig. 3.8 Purchased power from the pool market considering without and with DRP for scenario 50

equal to 158.0 MWh, which is increased to 160.90 MWh in the second case. In addition, the maximum procured power from the pool market is recorded as 22 and 21.8 MW at hours 15 and 13 for without and with considering DRP, respectively. At some periods, purchased power from the pool market is equal to zero which comes from the load profile of the large consumer after applying DRP. Obtained results show that participation of the pool market is slightly changed after applying the demand response program to reduce the total power procurement cost of the large consumer.

Figure 3.9 presents the procured power from the self-generating units in scenario 50. Total procured power from self-generating units is recorded as 316.5 MWh and 313.5 MWh for without and with considering DRP, respectively. In both cases, the maximum obtained power is reported as 25 MW at hour 20, while the minimum procured power is obtained as 9 MW at hour 12.

Procured power from bilateral contracts is depicted in Fig. 3.10. The total procured power from bilateral contracts is equal to 120 MWh, which is the same for without and with considering demand response program cases. The maximum total procured power from BCs is 7 MW, which is recorded between hours 20 and 24. Furthermore, the minimum procured power from the sum of 12 different bilateral contracts is equal to 4 MW, which is recorded during hours 1–19. Due to predetermined power prices of the bilateral contracts, obtained results are same with the deterministic model, which is detailed in Chap. 2. Note that the obtained

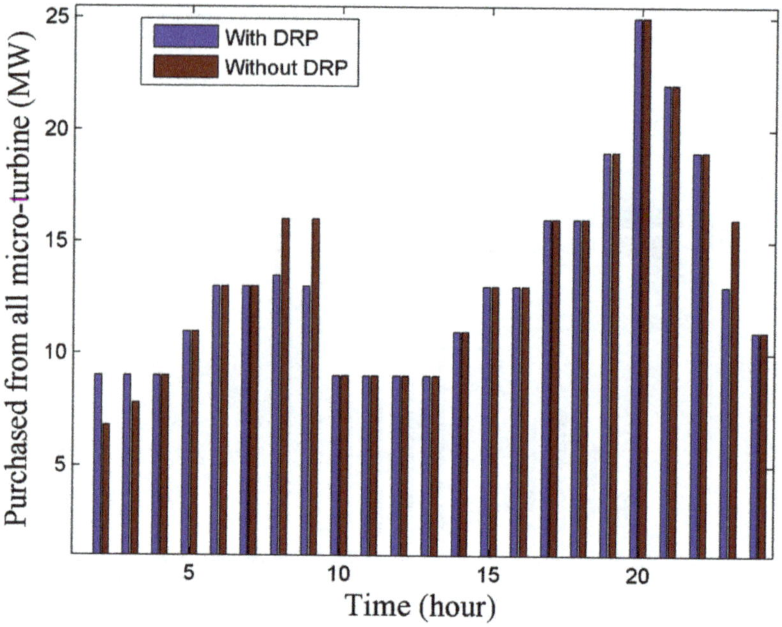

Fig. 3.9 Procured power from self-generating units considering with and without DRP for scenario 50

power of each hour in Fig. 3.10 is the sum of procured power from 12 bilateral contracts with different prices and periods.

The charging/discharging state of the battery storage system in scenario 50 is shown in Fig. 3.11 for the time horizon of the study considering without and with demand response programs. In this figure, the charged power is presented by positive values, while negative values present the discharged power. The ESS is discharged at hours 8, 19, 20, and 21 to supply some part of the required load of the large consumer. The battery storage system is charged during hours 1–3 and 11–13, which are the low-price periods, and discharged during 19–21, which are the high-price and demand periods.

Figure 3.12 presents the stored energy in the battery storage system in scenario 50 considering without and with demand response cases for the time horizon of the study. Total stored energy in the battery storage system is equal to 28.10 and 27.10 MW for without and with considering demand response programs, respectively. The maximum stored energy in the battery is recorded as 2 MWh, which is recorded at hours 13–18 for both cases. Furthermore, the minimum stored power is 0.2 MWh recorded at hours 21–24.

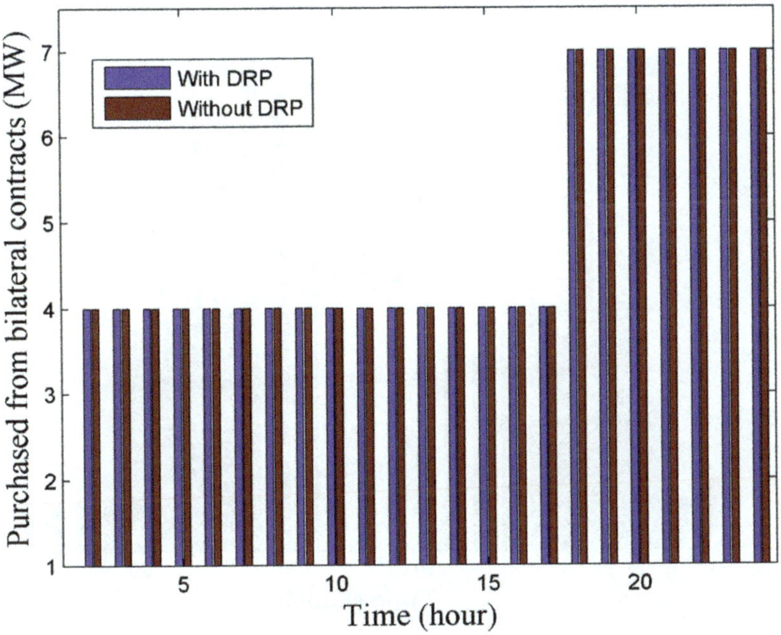

Fig. 3.10 Procured power from BCs considering with and without DRP

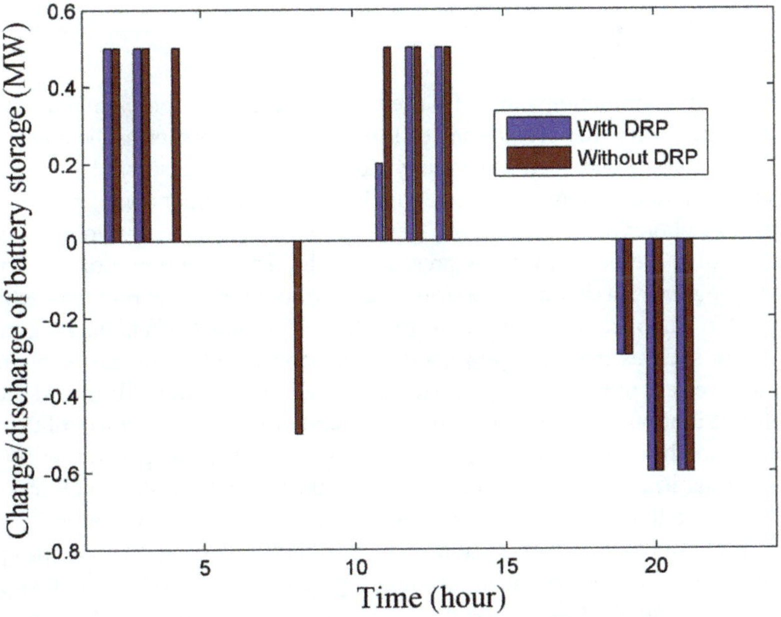

Fig. 3.11 Charging and discharging states of ESS considering with and without DRP

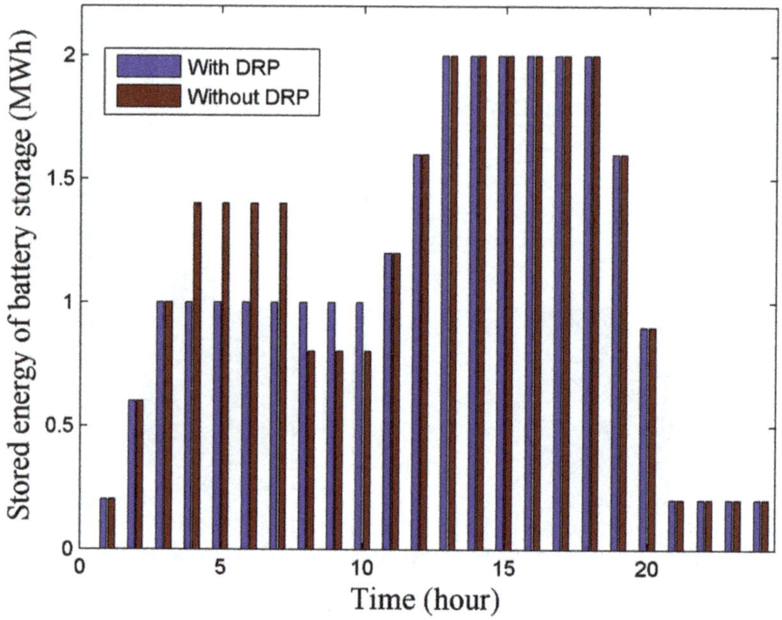

Fig. 3.12 Stored energy in the ESS considering with and without DRP

3.6 Conclusion

The stochastic programming method is used to solve the power procurement problem faced by a large consumer considering different uncertainties such as the power price in the pool market and load uncertainty. The considered case study in Chap. 2 including six alternative power sources including bilateral contracts, pool market, wind turbine, photovoltaic system, self-generating units, and energy storage systems is used to solve the problem in the presence of the time-of-use rate of the demand response program. Based on the deterministic formulation of the problem, which is introduced in Chap. 2, the stochastic formulation is provided. Then, uncertainties of the problem such as the load, power price, and power output of renewable-based energy sources are modeled using 100 discrete scenarios. It should be noted that the Weibull distribution is used to model the uncertainty of the output power of the wind turbines. Also, the output power of photovoltaic systems is model generating 100 scenarios for solar irradiance and temperature using the normal distribution. Finally, the power price and load scenarios are modeled using the normal distribution.

Two different cases as without and with considering demand response programs are assumed to solve the problem and investigate the impact of demand response programs. To do so, the time-of-use rate of demand response is applied to reduce the total power procurement cost of the large consumer. The problem is formulated as a mixed-integer linear programming, which is solved by the CPLEX solver under GAMS optimization software.

According to the obtained results, the cost of procured power from the pool market is \$12,827.042 for the first case. In the second case, it is reduced to \$11,241.042 by applying the demand response program. Comparing results shows about 12.36% reduction in the procured power cost from the pool market. The cost of purchased power from the bilateral contract is obtained as \$9322.080 for both cases. Due to predetermined prices of the bilateral contracts, the participation of these contracts has not changed. The operating cost of self-generating units, which is considered as the power procurement cost, is obtained as \$18,237.456 and \$17,964.255 for without and with considering demand response programs, respectively. Finally, the total power procurement cost, which is the sum of cost of procured power from different power procurement sources, is obtained as \$40,386.577 without considering the demand response program. By applying the time-of-use rate of demand response program, the total power procurement cost is reduced to \$38,527.377, which indicates that by applying the demand response programs, a 4.6% reduction is achieved in the total power procurement cost.

Obtained results in this section are presented for scenario 50 among 100 discrete scenarios, which are generated for modeling of the related uncertainties to the power procurement problem. The decision-maker has to choose the best strategy based on the obtained results.

References

1. H. Ebrahimian, S. Barmayoon, M. Mohammadi, N. Ghadimi, The price prediction for the energy market based on a new method. Econ. Res. Istraživanja **31**(1), 313–337 (2018)
2. V. Dua, N.A. Bozinis, E.N. Pistikopoulos, A multiparametric programming approach for mixed-integer quadratic engineering problems. Comput. Chem. Eng. **26**(4–5), 715–733 (2002)
3. J.R. Birge, F. Louveaux, *Introduction to Stochastic Programming* (Springer, New York, 2011)
4. P. Beraldi, A. Violi, G. Carrozzino, M.E. Bruni, A stochastic programming approach for the optimal management of aggregated distributed energy resources. Comput. Oper. Res. **96**, 200–212 (2018)
5. S. Xie et al., Multi-objective active distribution networks expansion planning by scenario-based stochastic programming considering uncertain and random weight of network. Appl. Energy **219**, 207–225 (2018)
6. F. Xie, Y. Huang, A multistage stochastic programming model for a multi-period strategic expansion of biofuel supply chain under evolving uncertainties. Transp. Res. Part E Logist. Transp. Rev. **111**, 130–148 (2018)
7. M.-C. Hu, S.-Y. Lu, Y.-H. Chen, Stochastic–multiobjective market equilibrium analysis of a demand response program in energy market under uncertainty. Appl. Energy **182**, 500–506 (2016)
8. M. Peker, A.S. Kocaman, B.Y. Kara, A two-stage stochastic programming approach for reliability constrained power system expansion planning. Int. J. Electr. Power Energy Syst. **103**, 458–469 (2018)

9. G. Schweickardt, V. Miranda, A two-stage planning and control model toward economically adapted power distribution systems using analytical hierarchy processes and fuzzy optimization. Int. J. Electr. Power Energy Syst. **31**(6), 277–284 (2009)
10. K. Marti, P. Kall (eds.), *Stochastic Programming*, vol 423 (Springer Berlin Heidelberg, Berlin, Heidelberg, 1995)
11. A.J. Conejo, M. Carrión, J.M. Morales, *Decision Making Under Uncertainty in Electricity Markets*, vol 153 (Springer US, Boston, MA, 2010)
12. Y. Liu, W. Wang, N. Ghadimi, Electricity load forecasting by an improved forecast engine for building level consumers. Energy **139**, 18–30 (2017)
13. H.A. Aalami, S. Nojavan, Energy storage system and demand response program effects on stochastic energy procurement of large consumers considering renewable generation. IET Gener. Transm. Distrib. **10**(1), 107–114 (2016)
14. M. Abbaspour, M. Satkin, B. Mohammadi-Ivatloo, F. Hoseinzadeh Lotfi, Y. Noorollahi, Optimal operation scheduling of wind power integrated with compressed air energy storage (CAES). Renew. Energy **51**, 53–59 (2013)
15. N. Ghadimi, A. Akbarimajd, H. Shayeghi, O. Abedinia, Application of a new hybrid forecast engine with feature selection algorithm in a power system. Int. J. Ambient Energy, 1–10 (2017)
16. D.T. Nguyen, L.B. Le, Optimal bidding strategy for microgrids considering renewable energy and building thermal dynamics. IEEE Trans. Smart Grid **5**(4), 1608–1620 (2014)
17. N. Ghadimi, A. Akbarimajd, H. Shayeghi, O. Abedinia, A new prediction model based on multi-block forecast engine in smart grid. J. Ambient Intell. Humaniz. Comput., 1–16 (2017)
18. D.K. Barrow, N. Kourentzes, Distributions of forecasting errors of forecast combinations: implications for inventory management. Int. J. Prod. Econ. **177**, 24–33 (2016)
19. S. Nojavan, H. Allah Aalami, Stochastic energy procurement of large electricity consumer considering photovoltaic, wind-turbine, micro-turbines, energy storage system in the presence of demand response program. Energy Convers. Manag. **103**, 1008–1018 (2015)
20. "CPLEX 12." [Online]. Available: https://www.gams.com/latest/docs/S_CPLEX.html. Accessed 15 Jul 2018
21. A. Brooke, D. Kendrick, A. Meeraus, GAMS user's guide. Redwood City (CA): The Scientific Press; 1990. Available: http://www.gams.com/docs/gams/GAMSUsersGuide.pdf

Chapter 4
Robust Optimization-Based Energy Procurement

Mahdi Mir, Noradin Ghadimi, Oveis Abedinia, and Sayed Ahmad Reza Shokrani

Abstract In this chapter, the robust optimization approach (ROA), which is one of the most popular uncertainty modeling methods, is used to solve the power procurement problem of a large consumer under the power price uncertainty considering 30% variation in the price. In contrast to stochastic optimization, ROA is rather a deterministic and set-based method. In addition, the robust optimization method investigates the effect of an uncertain parameter on the optimal result, which aims to reduce the sensitivity of the optimal result to the uncertain parameter. To solve the problem, the standard MILP formulation of proposed model based on deterministic formulation is provided and solved under CLPX solver in GAMS optimization program. The comparing obtained results with the deterministic case show that by increasing the power price in the pool market, the large consumer seeks to procure its required demand using other sources as self-generating units, which makes the consumer robust against the price volatility of the market.

M. Mir
Department of Electrical Engineering, Ferdowsi University of Mashhad, Mashhad, Iran

N. Ghadimi (✉)
Young Researchers and Elite Club, Ardabil Branch, Islamic Azad University, Ardabil, Iran
e-mail: noradin.ghadimi@iauardabil.ac.ir

O. Abedinia
Department of Electrical Engineering, Budapest University of Technology and Economics, Budapest, Hungary
e-mail: oveis.abedinia@vet.bme.hu

S. A. R. Shokrani
Department of Industrial Management, Faculty of Management, University of Tehran, Tehran, Iran
e-mail: shokrany@ut.ac.ir

S. Nojavan et al. (eds.), *Robust Energy Procurement of Large Electricity Consumers*,
https://doi.org/10.1007/978-3-030-03229-6_4

4.1 An Introduction to Robust Optimization Approach

Stochastic programming is used to model the uncertainty when the exact distribution of the required information of the uncertain parameter is well known or easily can be determined [1]. In the cases that appropriate distribution of the data is not available or stochastic programming is computationally impractical [2], the robust optimization approach (ROA) [3] can be considered to address the uncertainty in the system.

In contrast to stochastic optimization, which starts by considering probabilistic description for the uncertainty, ROA is rather a deterministic and set-based method [4]. In the ROA, the decision-maker can construct a feasible solution for any realization of the uncertainty in a given set. This method is categorized in the interval optimization methods and models the worst case of uncertainty parameter [5, 6]. In other words, ROA pursues to immunize the decision-maker from the worst realizations of outcomes, which makes it a conservative approach [7, 8]. In addition, the ROA has turned to a popular method considering its different features such as computational tractability [9].

In the ROA optimization method, the user should set the boundaries of uncertain parameter, and the critical value of the objective function would be calculated based on the predefined interval. More information about this method is provided in the following sections.

4.2 Differences of ROA and Stochastic Optimization Approach

The stochastic programming and other uncertainty modeling methods such as probabilistic and hybrid probabilistic-possibilistic methods are less preferred due to the high-required information about the uncertain parameters. The robust optimization method investigates the effect of an uncertain parameter on the optimal result, which aims to reduce the sensitivity of the optimal solution to the uncertain parameter. This approach can be considered as a substitution for the stochastic programming to address uncertainty in mathematical programming models. In comparison with other methods, ROA which is a risk management method has a low computing volume [10].

Both the stochastic programming and ROA are used to model and solve the uncertainty in the system [11–13]. The stochastic programming models each uncertain parameter as a random variable. The main difference between the stochastic programming and ROA comes from the fact that the uncertainty of various parameters such as solar radiation, wind speed, gas emission, and market price can be modeled simultaneously in the stochastic method [14], which is impossible in the ROA. In other words, as considered in Chap. 3, a set of possible scenarios, based on some simulation methods such as Monte Carlo, are generated to model uncertainties of each uncertain parameter including power price, load, solar radiation, temperature, and wind speed, which means that the global optimal results are not guaranteed by the obtained results. The ROA creates various strategies, and one of the provided strategies, according to the behavior of the uncertain parameter, can be used to guarantee the global optimal results [15].

4.3 Applications of ROA in the Power System Modeling

In recent years, ROA has gained significant attention to solve power system problems with uncertainty. The power market price, load demand, and power output of renewable sources are the main sources of uncertainty in the power system problems. The application of ROA in power system problems can be divided into the following cases: scheduling of microgrids, economic dispatch problem, and strategic bidding of electricity markets.

To integrate distributed energy resources (DER), microgrids (MGs) can be considered as a feasible solution. Obtaining the optimal scheduling of power generating units is the main motivation to provide the objective function. In a combined heat and power-based MG, ROA is proposed to get the scheduling of multiple generating units in [16], considering power price, heat, and power demand uncertainties. It should be noted that the power price uncertainty is modeled using the chance-constraint method. The optimal scheduling problem of MG is solved in grid-connected mode using the ROA to model the power output of renewable energy sources in [17]. The same problem is solved considering renewable energy sources in [18] using the ROA to calculate the allowable power output of a wind farm. A scenario-robust mixed-integer linear program is proposed in [19], to utilize ensemble weather forecasts to improve the performance of a hybrid microgrid including traditional power generation units and renewable sources.

Optimal bidding strategy is obtained in [20] for a price-taker producer by implementing the ROA considering maximum and minimum amount of power market price instead of the forecasted quantities. In [21], the ROA is utilized by an energy management system to model the uncertainties of power price in the market and output power of wind turbine. In [22], the robust mixed-integer linear programming model is proposed to help retailers to select a robust decision considering market price uncertainty. Meanwhile, stochastic programming is used to model the uncertainty of demand and variable climate condition. In this Ref, by using the introduced model, the optimal bidding and offering curves are derived for the problem.

In order to provide the unit commitment scheduling in the day-ahead power market, the unit commitment optimization model is introduced in [23]. The same problem is solved by implementing the ROA to model the uncertainties related to wind power generation in [24]. In [25], the ROA is used to solve the contingency-constrained unit commitment problem to get the optimal solution in the presence of simultaneous outage of generation units. The uncertainty of the marginal price, including price and reserve uncertainties, is considered to solve the unit commitment problem in a day-ahead power market by implementing the ROA in [26].

To minimize the operating cost of the system of a building energy system consisting of multi-chiller system, ice thermal energy storage, and solar power generation system, a two-stage adjustable robust formulation is proposed in [27].

Uncertainty of the load demand and wind power generation is modeled using the ROA in [28] for day-ahead scheduling of smart distribution networks under demand response programs. To do so, a tractable min–max–min cost model is proposed to obtain the optimal robust day-ahead scheduling of the smart distribution network.

The problem of transition network expansion planning is solved under the load and wind power uncertainties using a two-stage ROA in [29]. Finally, the same problem is solved using the ROA in [30], considering the load demand and renewable energy power generation uncertainties.

4.4 Robust Optimization-Based Power Procurement

In this section, the robust optimization method is used to solve the power procurement problem of the large consumer in the presence of power price uncertainty. To do so, first the formulation of the ROA is proposed. Then, the standard MILP formulation of the considered model is provided based on the deterministic formulation which is presented in Chap. 2.

4.4.1 Deterministic Model of the Case Study

In order to apply ROA method on the problem, the information of considered case study in Chap. 2 is used without considering the demand response program. In this case, the deterministic formulation of the problem which is provided in Eqs. (2.1)–(2.21) is used to implement the ROA on the problem.

4.4.2 Robust Optimization Model

The standard MILP formulation of proposed model based on Eqs. (2.1)–(2.21) can be defined as follows:

$$\text{Min} \sum_{t=1}^{n} c_t \cdot x_t \tag{4.1}$$

$$\text{s.t}: \sum_{t=1}^{n} a_{\mathrm{it}} \cdot x_t \leq b_i \quad \forall i = 1, \ldots. m \tag{4.2}$$

$$x_t \geq 0 \quad \forall t = 1, \ldots, n \tag{4.3}$$

$$x_t \in \{0, 1\} \quad \text{for some } t = 1, \ldots, n \tag{4.4}$$

where $x_t(\forall t)$ are the optimization variables. Each uncertain entry c_t is modeled as a random variable, independent, symmetric, and bounded (but with unknown distribution), which takes values in the interval $\left[c_t - \widehat{c}_t, c_t + \widehat{c}_t\right]$ with $\widehat{c}_t \geq 0$, being $\widehat{c}_t$ the corresponding deviation from the mean value of the coefficients c_t.

To adjust the robustness of the proposed method against the desired conservatism level, a control parameter, Γ, is defined for each constraint in order to formulate a robust mixed-integer linear program problem with uncertainty. Γ can take values in the interval [0, 1]. The problem (2.1)–(2.27) involves uncertain parameter in objective function coefficients c_t, particularly λ_h in the objective function.

The robust mixed-integer linear program problem corresponding to the problem (4.1)–(4.4) considering only uncertainty in objective function has the following form [3, 20]:

$$\text{Min} \sum_{t=1}^{n} c_t \cdot x_t + z \cdot \Gamma + \sum_{t=1}^{n} q_{\mathrm{t}} \tag{4.5}$$

$$\text{s.t} \sum_{t=1}^{n} a_{\mathrm{it}} \cdot x_t \leq b_i, \quad \forall i = 1, \ldots. m \tag{4.6}$$

$$z.\Gamma + q_t \geq \widehat{c}_t \cdot y_t \quad i = 1, \ldots, m \tag{4.7}$$

$$q_t \geq 0 \tag{4.8}$$

$$y_t \geq 0 \tag{4.9}$$

$$z \geq 0 \tag{4.10}$$

$$x_t \leq y_t \tag{4.11}$$

$$x_t \geq 0 \tag{4.12}$$

$$x_t \in \{0, 1\} \quad \text{for some } t = 1, \ldots, n \tag{4.13}$$

The optimization variables of the robust problem (4.5)–(4.13) are x_t, q_t, y_t, and z where z and q_t are dual variables of the original problem (4.1)–(4.4) used to take into account the range of variation of coefficients c_t, while y_t is an auxiliary variable used to obtain linear expressions. Also, Γ takes values in the interval [0, 1] if $\widehat{c}_t > 0$, while $\Gamma = 0$ if $\widehat{c}_t = 0$.

In model (2.1)–(2.21), the only uncertain parameter is the power price in the pool market at each hour, λ_t, which appears in the objective function. As discussed above, λ_t is modeled as random variable $\widehat{\lambda}_t$ which takes values in the interval $\left[\lambda_t - \widehat{\lambda}_t, \lambda_t + \widehat{\lambda}_t\right]$.

To get robust model in forms (4.5)–(4.13), the only uncertain parameter is in the objective function (2.1). Therefore, constraints (2.2)–(2.21) remain unchanged because there is not any uncertain parameter in them. It should be noted that demand response programs are not applied in this chapter, which in this case, the constraint (2.2) should be considered as provided in constraint (4.22). Considering (4.5)–(4.13), the robust model by considering uncertainty can be formulated as follows:

$$\text{Min OF} = z.\Gamma + \sum_{t=1}^{24} q_t + \sum_{l=1}^{B}\sum_{t=1}^{T} \lambda_{l,t} P_{l,t}^{\text{BC}} + \sum_{t=1}^{T} \left\{ \begin{array}{l} \lambda_t P_t^p + \sum_{j=1}^{N_j}\sum_{h=1}^{N_h} S_{j,h}^{\text{DG}} P_{j,h,t} \\ +C_t^B \times \left(\chi \times P_t^{\text{charge}} + \frac{P_t^{\text{disc}}}{\eta}\right) \\ +C_t^{\text{wind}} \times P_t^{\text{wind}} + C_t^{\text{PV}} \times P_t^{\text{PV}} \end{array} \right\} \tag{4.14}$$

Subject to:

$$z + q_t \geq \widehat{\lambda}_t . y_t \quad \forall t \mid \widehat{\lambda}_t > 0 \tag{4.15}$$

$$q_t \geq 0 \quad \forall t \mid \widehat{\lambda}_t > 0 \tag{4.16}$$

$$y_h \geq 0 \tag{4.17}$$

$$z \geq 0 \tag{4.18}$$

$$P_t^p \leq y_t \tag{4.19}$$

$$q_t \geq 0 \quad \forall t \mid \widehat{\lambda}_t > 0 \tag{4.20}$$

$$\text{Constraints}\,(2.2) - (2.21) \tag{4.21}$$

The robustness parameter, Γ, takes values in the interval [0, 1] if there is uncertainty in electricity power price. Constraints (4.15)–(4.20) are imposed by the robust model in addition to previous constraints.

The constraints (2.2)–(2.21) are presented in the following:

$$\sum_{l=1}^{B} P_{l,t}^{\text{BC}} + P_t^p + \sum_{j=1}^{N_j}\sum_{h=1}^{N_h} P_{j,h,t} + P_t^{\text{wind}} + P_t^{\text{PV}} + P_t^{\text{disc}} = \text{load}_t^0 + P_t^{\text{charge}} \tag{4.22}$$

$$P_{l,t}^{\min} s_l < P_{l,t}^{\text{BC}} < P_{l,t}^{\max} s_l; \quad \forall l, t \tag{4.23}$$

$$P_t^{\mathrm{BC}} = \sum_{l=1}^{N_l} P_{l,t}^{\mathrm{BC}}; \quad \forall t \tag{4.24}$$

$$\mathrm{Cost} = \sum_{t=1}^{T} \sum_{j=1}^{N_j} \sum_{h=1}^{N_h} S_{j,h}^{\mathrm{DG}} P_{j,h,t} \tag{4.25}$$

$$0 \le P_{j,h,t} \le \left(P_{j,h}^{\mathrm{MAX}} - P_{j,h-1}^{\mathrm{MAX}}\right) \times U_{j,t}; \quad \forall j,h,t \tag{4.26}$$

$$0 \le P_{j,1,t} \le P_{j,1}^{\mathrm{MAX}} \times U_{j,t}; \quad \forall j,t \tag{4.27}$$

$$\sum_{h=1}^{N_h} P_{j,h,t} - \sum_{h=1}^{N_h} P_{j,h,t-1} \le R_j^{\mathrm{up}} \times U_{j,t}; \quad \forall j,t \tag{4.28}$$

$$\sum_{h=1}^{N_h} P_{j,h,t-1} - \sum_{h=1}^{N_h} P_{j,h,t} \le R_j^{\mathrm{down}} \times U_{j,t-1}; \quad \forall j,t \tag{4.29}$$

$$\begin{array}{l} U_{j,t} - U_{j,t-1} \le U_{j,t+\mathrm{Up}_{j,i}}; \quad \forall j, \forall t, \forall i \\ \mathrm{U}_{j,t-1} - \mathrm{U}_{j,t} \le 1 - \mathrm{U}_{j,t+\mathrm{Dn}_{j,i}}; \quad \forall j, \forall t, \forall i \end{array} \tag{4.30}$$

$$\begin{array}{l} \mathrm{Up}_{j,i} = \left\{ \begin{array}{ll} i & i \le \mathrm{MUT}_j \\ 0 & i \succ \mathrm{MUT}_j \end{array} \right\} \\ \mathrm{Dn}_{j,i} = \left\{ \begin{array}{ll} i & i \le \mathrm{MDT}_j \\ 0 & i \succ \mathrm{MDT}_j \end{array} \right\} \end{array} \tag{4.31}$$

$$P_{t,s}^{\mathrm{wind,\,max}} = \begin{cases} 0 & V_{t,s}^{w} < V_{\mathrm{ci}} \\ p_r \times \left(\frac{V_{t,s}^{w} - V_{\mathrm{ci}}}{V_r - V_{\mathrm{ci}}}\right)^3 & V_{ci} < V_{t,s}^{w} < V_{\mathrm{cr}} \\ p_r & V_r < V_{t,s}^{w} < V_{c0} \\ 0 & V_{t,s}^{w} > V_{c0} \end{cases} \tag{4.32}$$

$$P_t^{\mathrm{wind}} \le P_t^{\mathrm{wind,\,max}}; \quad \forall t \tag{4.33}$$

$$P_t^{M,\max} = \frac{G_t^a}{G_{a_0}} \times \left\{ P_{\mathrm{Max},0}^{M} + \mu_{P\max} \times \left(T_{t,}^{a} + G_t^a \times \frac{\mathrm{NOCT} - 20}{800} - T_{M,0} \right) \right\} \tag{4.34}$$

$$P_t^{\mathrm{PV}} \le P_t^{M,\max}; \quad \forall t \tag{4.35}$$

$$X_{t_0}^{b} = X_0^{b} \tag{4.36}$$

$$P_t^{\mathrm{charge}} \le P_{\mathrm{charge}}^{\max} \times U_t^{\mathrm{charge}}; \quad \forall t \tag{4.37}$$

$$P_t^{\mathrm{disc}} \le P_{\mathrm{disc}}^{\max} \times U_t^{\mathrm{disc}}; \quad \forall t \tag{4.38}$$

$$X_b^{\min} \le X_t^{b} \le X_b^{\max}; \quad \forall t \tag{4.39}$$

$$U_t^{\text{charge}} + U_t^{\text{disc}} \leq 1; \quad \forall t \tag{4.40}$$

$$X_t^b = X_{t-1}^b + \chi \times P_t^{\text{charge}} - \frac{P_t^{\text{disc}}}{\eta}; \quad \forall t \tag{4.41}$$

In the constraint (4.22)–(4.41), $P_{l,t}^{\text{BC}}$ is the procured power from the BCs with predetermined price $\lambda_{l,t}$; P_t^p is the cost of energy procurement from the pool market with the market price λ_t; $P_{j,h,t}$ is the cost of produced power by the jth unit of the self-generation units; C_t^B is the operating cost of using battery storage, with charging/discharging efficiency χ/η and charging/discharging power of battery storage (MW) $P_t^{\text{charge}}/P_t^{\text{disc}}$; C_t^{wind} is the cost of produced power from the wind turbine; C_t^{PV} is the cost of produced power P_t^{PV} from the PV system.

Also, load_t^0 is the initial load, $P_{l,t}^{\min}/P_{l,t}^{\max}$ is the minimum/maximum capacity relation, P_t^{BC} is the total procured power from the bilateral contracts, $P_{j,h}^{\text{MAX}}$ is the power output of block h of jth unit of self-generating units, $R_j^{\text{up}}, R_j^{\text{down}}$ is ramp up/down rate limit of j unit, MUT_j,MDT_j is the minimum up-/downtime of self-generation units j, V_t^w is the wind speed, V_r is the rated speed, V_{ci} is the cut-in speed, V_{c0} is the cutout wind speed, $P_t^{M,\max}$ is the maximum power at any temperature T_t^a, G_t^a is irradiation, $P_{\text{Max},0}^M$ is the maximum power, and NOCT is the normal operating cell. For more information about the constraints (4.22)–(4.41), please refer to Chap. 2.

4.4.3 The Algorithm of Robust Optimization Problem

The power procurement problem faced by a large consumer considering the power price uncertainty can be modeled and solved using ROA by taking the following steps:

- Step 0: Set market price deviation due to uncertainty $\widehat{\lambda}_t = 0.3 \times \lambda_t$ and $\Gamma = 0$ (the deterministic model is obtained).
- Step 1: Set $\Gamma = G^k$, where G^k is a coefficient that uses increasing values within $[0, 1]$ and k is the counter of iteration.
- Step 2: Robust mixed-integer linear programming optimization (4.14)–(4.21) is solved to obtain the procured power from alternative resources as pool market, bilateral contracts, self-generating units, etc. at the iteration k.
- Step 3: In order to cover all ranges of coefficient G^k, steps 1 and 2 should be repeated iteratively (categorized by k) as illustrated in Fig. 4.1.
- Step 4: Display the obtained results, which present procured power from different power sources ($P_t^p, P_{l,t}^{\text{BC}}, P_{j,h,t}, P_t^{\text{charge}}, P_t^{\text{disc}}, P_t^{\text{wind}}, P_t^{\text{PV}}$), for each period and robustness parameter value, Γ^k.

Figure 4.1 is presented to clarify the proposed algorithm to solve the power procurement problem using the robust mixed-integer linear programming optimization.

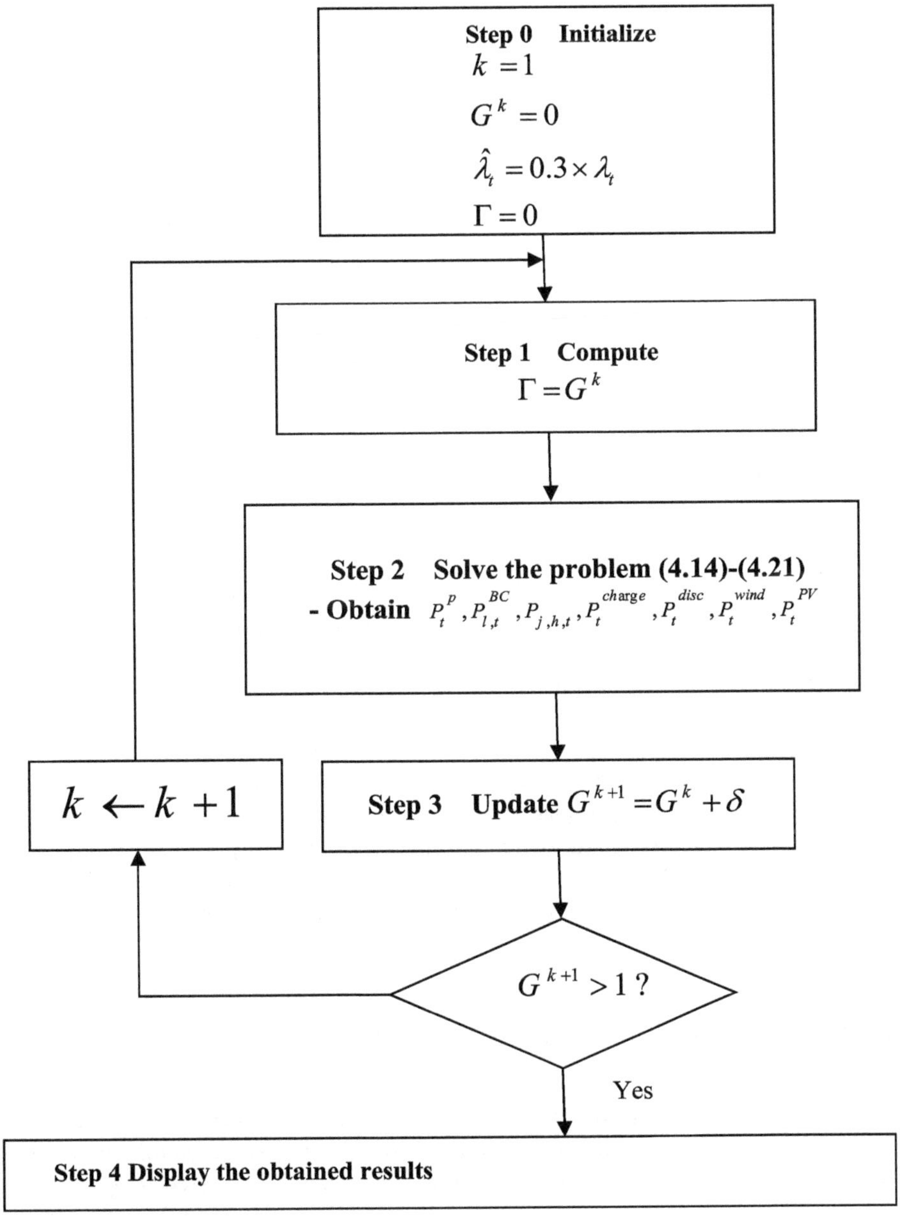

Fig. 4.1 Flowchart of the robust optimization approach

4.5 Case Study

By using the considered case study in Chap. 2, the power procurement problem faced by a large consumer is solved considering the power price uncertainty in the power market. In order to show the impact of the power price uncertainty, obtained results are compared with the deterministic case, which is presented in Chap. 2.

4.6 Results and Discussion

In this section, obtained results for the deterministic and robust optimization methods are presented. The studied case in Chap. 2 without considering the demand response program is used to show the impact of the proposed method. Deterministic mode formulation is provided in (2.1)–(2.21). Furthermore, the robust optimization mode is formulated in Eqs. (4.14)–(4.21).

4.6.1 Deterministic Case Results

In order to get deterministic case results, the problem is solved using Eqs. (2.1)–(2.21); the total power procurement cost of the large consumer is equal to \$40,253.252 without considering DRP. The cost of procured power from the pool market is equal to \$13,168.363. Furthermore, the cost of procured power from the bilateral contracts is obtained as \$9322.080. Finally, the procurement cost of self-generating units is equal to \$17,762.808. The obtained results are detailed in Chap. 2.

The results show that the consumer should procure 3.5% of the required load from the wind turbine, 1.38% from the photovoltaic system, 25.32% from the pool market, 50.1% from the self-generating units, and 19.36% from the bilateral contracts. The rest of the required load is provided by the energy storage system. The abovementioned numbers indicate that most of the required demand are supplied by the internal power procurement resources rather than by the pool market.

4.6.2 Robust Optimization Case Results

As said before, to get robust optimization case results, the problem is solved using Eqs. (4.14)–(4.21). The problem is solved considering 30% variation for power price in the pool market. The problem is formulated as robust mixed-integer linear programming and solved under CPLX solver [31] in the GAMS optimization program [32].

Figure 4.2 presents the power price assuming 30% variation for the power price for the time horizon of the study. In Fig. 4.2, the power price in the deterministic case is shown as expected, the power price with the 30% increase as the maximum, and the power price with the 30% reduction as the minimum.

Figure 4.3 illustrates the power procurement cost from the available resources for different values of the robustness parameter Γ. The ROA model is solved by taking 0.1 step size for the robustness parameter Γ, considering 11 iterations, and starting from zero to one. It should be denoted that by considering zero for the robustness parameter, the deterministic case results is obtained. Also, by assuming one for the robustness parameter, the worst-case results are obtained. Based on obtained results, which is shown in Fig. 4.3, if the consumer desires to get higher robustness against the pool price increase, it should pay a higher cost, and inversely, if the consumer pays a high cost, which presents its procurement strategy, it is highly robust and more risk-averse. Although, in this particular case, the high capacity of self-generation units, wind turbine, photovoltaic system, and bilateral contracts provides about 75% of the required demand, the large consumer is less sensitive against the pool market increase.

According to Fig. 4.3, by increasing the robustness parameter which shows the power price in the market, obtained energy from the pool market is reduced, while procured power from the self-generating units is increased which makes power procurement process less sensitive to power price increase in the market. In addition,

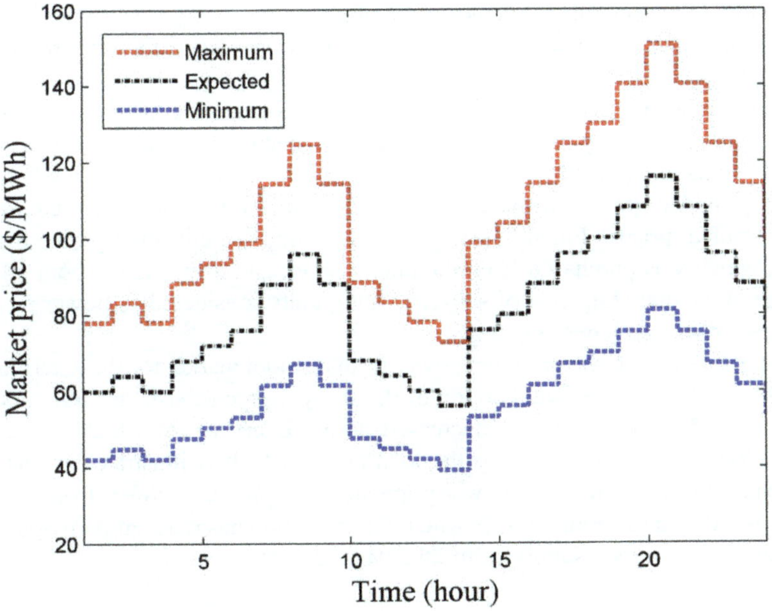

Fig. 4.2 The power price in the market considering 30% variation

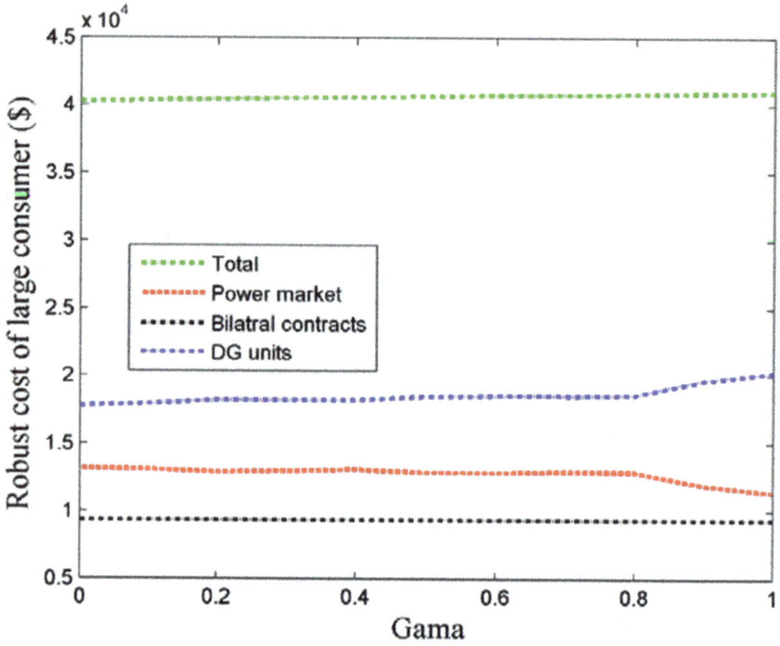

Fig. 4.3 Procurement cost of alternative power sources versus the robustness parameter

owing to the predetermined prices of the bilateral contracts obtained, power is not changed by increasing the robustness parameter.

Obtained results for deterministic and the worst case in the ROA are presented in Table 4.1.

According to provided results in Table 4.1, by considering 30% increase of the power price, procured power from the market is decreased which results to about 13.45% reduction in the power procurement cost from the pool market. As discussed before, provided power from the bilateral contracts has not changed due to predetermined prices. Finally, self-generating units should supply the reduced power, which was purchased from the pool market, and it results to about 13.89% increase in the operating cost of self-generating units considered as power procurement cost from self-generating units.

Figure 4.4 presents the procured power from the pool market for the deterministic and robust modes in the time horizon of the study. In the robust case, the procured power from the pool market is decreased during hours 14–24 which is the peak load period based on considering the load curve, which is illustrated in Fig. 2.7. The results show that by 30% power price increase, procured power from the pool market is reduced to about 15.67% which this reduction has reduced participation of the pool market in load supply from 25.32% to 21.35%.

Table 4.1 Simulation results of robust optimization approach versus deterministic approach

Different costs	Deterministic case	Robust optimization case
Procurement cost from pool market ($)	13,168.363	11,397.631
Procurement cost from BCs ($)	9322.080	9322.080
Procurement cost from SG units ($)	17,762.808	20,229.000
Total procurement cost ($)	40,253.252	40,948.710

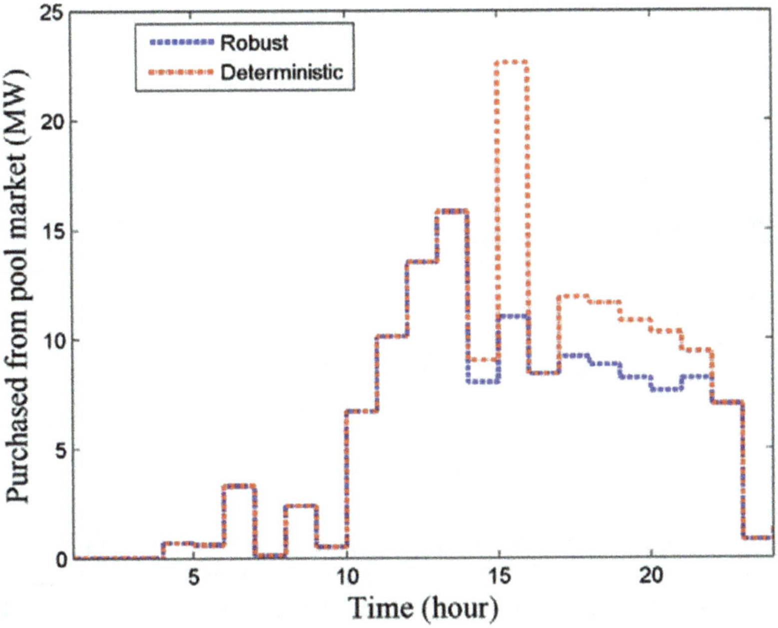

Fig. 4.4 Purchased power from the pool market

Procured power from the self-generating units is presented in Fig. 4.5 for deterministic and robust cases. As was expected, procured power from the self-generating units is increased to compensate the reduced purchased power from the pool market. In the worst case, considering 30% of the power price increase, the operating cost of the self-generating units is increased to about $2466.192, which shows about 13.89% increase in comparison with the deterministic case results. In addition, it should be denoted that procured power from the self-generating units is increased to about 7.24% by comparing deterministic case results, which indicates that in the robust case, 54.05% of the required load by the large consumer is provided by these units.

Purchased power from the bilateral contracts is depicted in Fig. 4.6 considering 24 h and the robust and deterministic cases. As was expected, due to fixed and

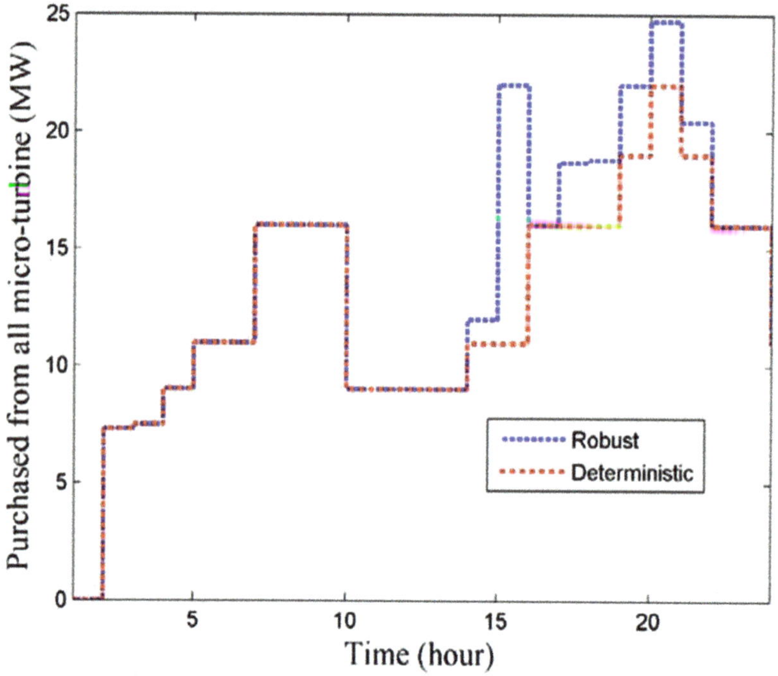

Fig. 4.5 Procured power from the self-generating units

predetermined price of the bilateral contracts, procured power from this source has not change in the presence of the power price uncertainty. It should be noted that the same results is obtained in Chap. 3, which the stochastic programming approach was used to model the power price uncertainty. According to Table 4.1, the total cost of purchased power from the bilateral contracts is obtained as $9322.080 for both cases. In addition, the large consumer procures 19.36% of its required load from the bilateral contracts in the robust case.

Figure 4.7 presents the charging and discharging states of the battery for 24 h. According to Fig. 4.7, participation of the battery during off-peak period is equal for the deterministic and robust cases, while during peak periods, more power is taken from the battery in the robust case. It should be noted that required power form of the battery storage system is negligible.

Finally, stored energy in the battery storage system is illustrated in Fig. 4.8. During the peak load period, by increasing the power price in the market, stored energy in the battery storage system is discharged to provide some part of the load. In addition, as discussed in Chap. 1, it is obvious that during off-peak load periods, the battery is charged and stored power is used during high-price periods.

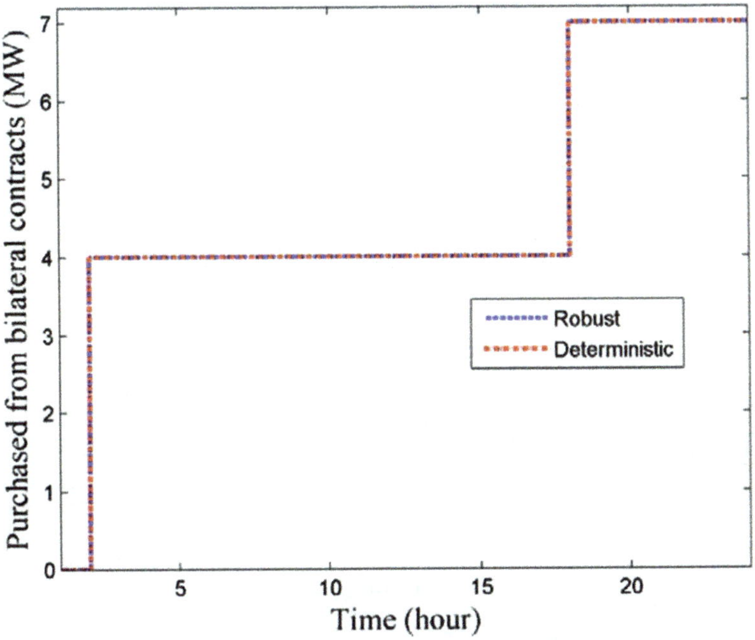

Fig. 4.6 Purchased power from the bilateral contracts for the robust and deterministic case results

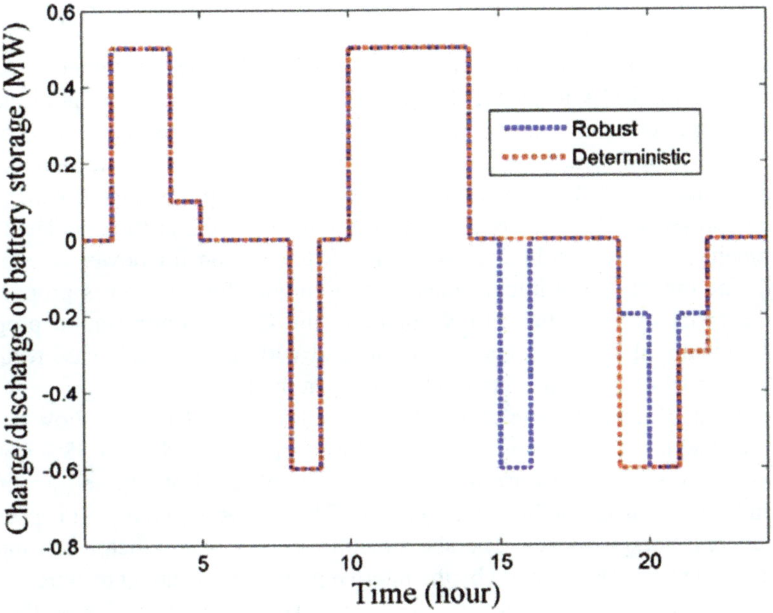

Fig. 4.7 Charge/discharge of the battery

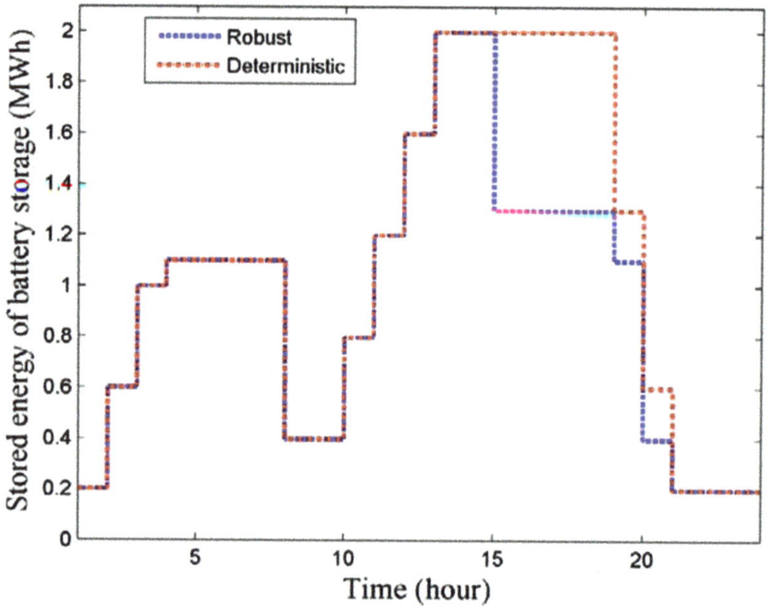

Fig. 4.8 Stored energy in the battery storage system

4.7 Conclusion

In this chapter, the power procurement problem of large consumer is solved implementing the ROA to model the power price uncertainty in the pool market. The ROA is preferred over other uncertainty modeling methods due to less computational burden and reliable results. In addition, the robust optimization method investigates the effect of an uncertain parameter on the optimal result, which aims to reduce the sensitivity of the optimal result to the uncertain parameter. The power procurement problem is solved considering 30% variation for power price in the market. The robust mixed-integer linear programming formulation is proposed to solve the problem under the CPLX solver in the GAMS optimization program. Obtained results of the worst case are compared with deterministic case results to show the impact of the power price uncertainty in the market.

The results of deterministic case, which was detailed in Chap. 2, show that the large consumer procure 3.5% of the required load from wind turbine, 1.38% from the photovoltaic system, 25.32% from the pool market, 50.1% from the self-generating units, and 19.36% from the bilateral contracts. The rest of required load is provided by the energy storage system. The abovementioned numbers indicate that most of the required demand are supplied by the internal power procurement resources rather than by the pool market. In this case, the total power procurement cost of the large consumer is equal to $40,253.252 without considering DRP. The cost of procured power from the pool market is equal to \$13,168.363. Furthermore, the cost of procured power from the bilateral contracts is obtained as \$9322.080. Finally, procurement cost from self-generating units is equal to \$17,762.808.

As said before, in the robust case, the problem is solved considering 30% variation in power price in the pool market. The ROA model is solved by taking 0.1 step size for the robustness parameter Γ, considering 11 iterations, and starting from zero to one. In the first iteration, the deterministic case results are obtained considering zero for the robustness parameter. Also, by assuming one for the robustness parameter, the worst-case results are obtained. Based on the obtained results, to get the higher robustness against the pool price increase, the higher power procurement should be paid by the consumer. In the assumed case study, the high capacity of self-generation units, wind turbine, photovoltaic system, and bilateral contracts, which provide about 75% of the required demand, has resulted in less sensitivity against the pool market increase.

By increasing the robustness parameter which shows the power price in the market, obtained energy from the pool market is reduced, while procured power from the self-generating units is increased which makes power procurement process less sensitive to power price increase in the market. In addition, owing to the predetermined prices of the bilateral contracts, obtained power is not changed by increasing the robustness parameter.

By considering the increase of the power price in the pool market, procured power from the market is decreased which results to about 13.45% reduction in the power procurement cost from the pool market. Provided power from the bilateral contracts has not changed due to predetermined prices. Finally, self-generating units should supply the reduced procured power from the pool market, which results to about 13.89% increase in the operating cost of self-generating units.

References

1. H.A. Bagal, Y.N. Soltanabad, M. Dadjuo, K. Wakil, N. Ghadimi, Risk-assessment of photovoltaic-wind-battery-grid based large industrial consumer using information gap decision theory. Sol. Energy **169**, 343–352 (2018)
2. N. Ghadimi, A new hybrid algorithm based on optimal fuzzy controller in multimachine power system. Complexity **21**(1), 78–93 (2015)
3. D. Bertsimas, M. Sim, The price of robustness. Oper. Res. **52**(1), 35–53 (2004)
4. D. Bertsimas, D.B. Brown, C. Caramanis, Theory and applications of robust optimization. SIAM Rev. **53**(3), 464–501 (2011)
5. S. Nojavan, B. Mohammadi-Ivatloo, K. Zare, Robust optimization based price-taker retailer bidding strategy under pool market price uncertainty. Int. J. Electr. Power Energy Syst. **73**, 955–963 (2015)
6. I. Ahmadian, O. Abedinia, N. Ghadimi, Fuzzy stochastic long-term model with consideration of uncertainties for deployment of distributed energy resources using interactive honey bee mating optimization. Front. Energy **8**(4), 412–425 (2014)
7. Z. Haider, H. Charkhgard, C. Kwon, A robust optimization approach for solving problems in conservation planning. Ecol. Model. **368**, 288–297 (2018)

8. A. Noruzi, T. Banki, O. Abedinia, N. Ghadimi, A new method for probabilistic assessments in power systems, combining Monte Carlo and stochastic-algebraic methods. Complexity **21**(2), 100–110 (2015)
9. Y. Zhang, J. Tang, A robust optimization approach for itinerary planning with deadline. Transp. Res. E Logist. Transp. Rev. **113**, 56–74 (2018)
10. R.M. Lima, A.Q. Novais, A.J. Conejo, Weekly self-scheduling, forward contracting, and pool involvement for an electricity producer. An adaptive robust optimization approach. Eur. J. Oper. Res. **240**(2), 457–475 (2015)
11. Z. Geng, A.J. Conejo, Q. Chen, C. Kang, Power generation scheduling considering stochastic emission limits. Int. J. Electr. Power Energy Syst. **95**, 374–383 (2018)
12. S. Nojavan, H. Allah Aalami, Stochastic energy procurement of large electricity consumer considering photovoltaic, wind-turbine, micro-turbines, energy storage system in the presence of demand response program. Energy Convers. Manag. **103**, 1008–1018 (2015)
13. N. Sayyad, B. Mohammadi-Ivatloo, K. Zare, Optimal bidding strategy of electricity retailers using robust optimisation approach considering time-of-use rate demand response programs under market price uncertainties. IET Gener. Transm. Distrib. **9**(4), 328–338 (2015)
14. A.J. Conejo, M. Carrión, J.M. Morales, *Decision Making Under Uncertainty in Electricity Markets*, vol 153 (Springer, Boston, 2010)
15. A. Najafi-Ghalelou, S. Nojavan, K. Zare, Heating and power hub models for robust performance of smart building using information gap decision theory. Int. J. Electr. Power Energy Syst. **98**, 23–35 (2018)
16. R. Wang, P. Wang, G. Xiao, A robust optimization approach for energy generation scheduling in microgrids. Energy Convers. Manag. **106**, 597–607 (2015)
17. Y. Zhang, N. Gatsis, G.B. Giannakis, Robust energy management for microgrids with high-penetration renewables. IEEE Trans. Sustain. Energy **4**(4), 944–953 (2013)
18. W. Wu, J. Chen, B. Zhang, H. Sun, A robust wind power optimization method for look-ahead power dispatch. IEEE Trans. Sustain. Energy **5**(2), 507–515 (2014)
19. E. Craparo, M. Karatas, D.I. Singham, A robust optimization approach to hybrid microgrid operation using ensemble weather forecasts. Appl. Energy **201**, 135–147 (2017)
20. L. Baringo, A.J. Conejo, Offering strategy via robust optimization. IEEE Trans. Power Syst. **26** (3), 1418–1425 (2011)
21. M. Rahimiyan, L. Baringo, Strategic bidding for a virtual power plant in the day-ahead and real-time markets: a price-taker robust optimization approach. IEEE Trans. Power Syst. **31**(4), 2676–2687 (2016)
22. S. Nojavan, K. Zare, B. Mohammadi-Ivatloo, Robust bidding and offering strategies of electricity retailer under multi-tariff pricing. Energy Econ. **68**, 359–372 (2017)
23. R. Jiang, J. Wang, Y. Guan, Robust unit commitment with wind power and pumped storage hydro. IEEE Trans. Power Syst. **27**(2), 800–810 (2012)
24. P. Xiong, P. Jirutitijaroen, C. Singh, A distributionally robust optimization model for unit commitment considering uncertain wind power generation. IEEE Trans. Power Syst. **32**(1), 39–49 (2017)
25. A. Street, F. Oliveira, J.M. Arroyo, Contingency-constrained unit commitment with $n - K$ security criterion: a robust optimization approach. IEEE Trans. Power Syst. **26**(3), 1581–1590 (2011)
26. H. Ye, Y. Ge, M. Shahidehpour, Z. Li, Uncertainty marginal price, transmission reserve, and day-ahead market clearing with robust unit commitment. IEEE Trans. Power Syst. **32**(3), 1782–1795 (2017)
27. C. Wang, B. Jiao, L. Guo, Z. Tian, J. Niu, S. Li, Robust scheduling of building energy system under uncertainty. Appl. Energy **167**, 366–376 (2016)
28. M. Mazidi, H. Monsef, P. Siano, Robust day-ahead scheduling of smart distribution networks considering demand response programs. Appl. Energy **178**, 929–942 (2016)
29. R.A. Jabr, Robust transmission network expansion planning with uncertain renewable generation and loads. IEEE Trans. Power Syst. **28**(4), 4558–4567 (2013)

30. H. Yu, C.Y. Chung, K.P. Wong, Robust transmission network expansion planning method with Taguchi's orthogonal array testing. IEEE Trans. Power Syst. **26**(3), 1573–1580 (2011)
31. CPLEX 12. [Online]. Available: https://www.gams.com/latest/docs/S_CPLEX.html. Accessed 15 Jul 2018
32. A. Brooke, D. Kendrick, A. Meeraus, R. Raman, R.E. Rosenthal, *GAMS A User's Guide Introduction 1* (GAMS Development Corporation, Washington, DC, 1998)

Chapter 5
Risk-Based Energy Procurement via IGDT

Moeid Dehghanpour Farashah

Abstract In this chapter, the power procurement problem of a large consumer under power price uncertainty is solved using information gap decision theory (IGDT). In contrast to other uncertainty modeling methods, the IGDT method makes it possible to model the positive and negative impacts of uncertainty through different strategies. To do so, the risk of the power procurement process of a large consumer is assessed considering the opportunity and robustness functions of the IGDT method, and three different strategies, risk-averse, risk-neutral, and risk-taker, are derived. It should be noted that the problem is formulated as mixed integer nonlinear programming and solved using GAMS programming software.

Based on the obtained results, in the risk-averse strategy, the large consumer is 12.16% robust against the power price uncertainty without implementing demand response program (DRP) by paying $42,000 to meet electricity demand. In addition, considering $40,000 as the procurement cost, 11.4% robustness is obtained against the power price increase in the pool market due to implement TOU-DRP. Also, in the risk-taker strategy, for a 15% power price reduction in the market, the large consumer's costs will be $37,250 and $35,700 in cases without and with DRP, respectively, which means DRP will reduce procurement costs by about 4.2%.

5.1 Introduction

In this chapter, information gap decision theory (IGDT) is used to solve the power procurement problem in an uncertain environment. The IGDT method, which is based on quantitative models, provides numerical assessment for the decision-making process. Uncertainty in a system might have a negative impact leading to high costs or a positive impact leading to low costs (windfall benefits). In the power procurement problem of a large consumer, power market price increases will have negative effects on the problem while a reduction in price will lead to lower procurement costs. Therefore, IGDT helps decision makers make the optimal

M. D. Farashah (✉)
3 Med Group, Tehran, Iran

S. Nojavan et al. (eds.), *Robust Energy Procurement of Large Electricity Consumers*,
https://doi.org/10.1007/978-3-030-03229-6_5

decision in light of developing strategies, evolving preferences, and available options. To address the conflicting aspects of uncertainty, opportunity and robustness functions are proposed in IGDT method in order to evaluate the positive and negative aspects of uncertainty.

In contrast to conventional uncertainty modeling methods such as stochastic programming, IGDT requires little information about the uncertainty structure, which significantly reduces the computational burden. In addition, in IGDT, the uncertainty parameter is chosen based on minimizing opportunity or maximizing robustness. Furthermore, the main difference between IGDT and other uncertainty modeling methods lies in the fact that in IGDT, the uncertainty parameter is initially undetermined. More information about IGDT is provided in the section 5.2.

To evaluate the uncertainty in a system using IGDT, different models are introduced: an energy-bound model, spatial analogy to info-gap uncertainty, envelope-bound model, Minkowski-norm model, slope-bound model, Fourier-bound model, hybrid info-gap model, combined info-gap model, nonconvex info-gap model, discrete info-gap model, and fractional error model. Each of these models is detailed in [1]. It should be noted that the fractional error model is used to model the uncertainty in the power procurement problem. In the following subsections, a comparison of IGDT and the most commonly used methods for uncertainty modeling, stochastic programming and robust optimization, is provided.

5.1.1 Comparison of Stochastic Programming and IGDT

The stochastic programming method is detailed and used to solve the power procurement problem in Chap. 3. Modeling of various types of uncertainty in a system, such as those related to market price, load demand, solar radiation, and wind speed, can be easily implemented using stochastic programming [2], while this is impossible in the IGDT method. This can be reckoned as the main difference between the IGDT and stochastic programming methods, although it is not considered a disadvantage for IGDT [3]. As discussed in Chap. 4, stochastic programming does not guarantee globally optimal results in terms of power procurement cost, while in IGDT various strategies are based on the behavior of the uncertainty parameter to guarantee predefined power procurement costs by choosing one of the developed strategies.

5.1.2 Comparison of Robust Optimization and IGDT

Robust optimization can be used to solve the power procurement problem, as detailed in Chap. 4. There are similarities and differences between the robust optimization and IGDT methods, both categorized as interval optimization methods. Considering the worst case of the uncertainty parameter is one of the similarities of the methods, while the most important difference lies in the input parameters of the methods, in which the desired amount of objective function is treated as an input in IGDT, while the maximum and minimum values of the uncertainty parameter should serve as input in the robust optimization method. In addition, IGDT analyzes the best case of uncertainty parameters [4, 5], which cannot be modeled using robust optimization. For more information about this topic, please refer to [6].

5.1.3 Application of IGDT Method

In the literature, IGDT is widely used to model uncertainty in power system problems, which can be divided into price uncertainty, load uncertainty, and power output of renewable energy sources.

In [7], price uncertainty in the pool market is modeled using the IGDT method to arrive at the optimal bidding strategy of a generation station in a day-ahead power market. It should be noted that in this reference, modified particle swarm optimization (MPSO) is considered to solve the optimization problem. The same problem is solved in [8], without using MPSO. To arrive at the optimal offering strategy for a large consumer, the IGDT method is implemented in [5] to model power price uncertainty under a demand response program. In [9], IGDT is used to forecast the power price in a market by a generator obtaining an optimal bidding curve to participate in a day-ahead electricity energy market. The power procurement problem of a large consumer is solved using IGDT to model power price uncertainty in [10]. Utilizing IGDT, demand bidding is constructed for a large consumer in [11]. Price uncertainty is modeled via IGDT in [12] to arrive at an optimal bidding strategy for industrial consumers considering cogeneration facilities to participate in a day-ahead electricity market. For merchant energy storage facilities, the optimal offer and bid strategies are derived considering power price uncertainty in the market, modeled using IGDT, in [13]. For robust scheduling of a smart apartment building, IGDT is used in [14] in the presence of a solar thermal storage system and price uncertainty. In [15], to manage the revenue risk of an electric vehicle aggregator, IGDT is used to model the uncertainty of electricity prices.

The uncertainty of the output power of a wind turbine is modeled using IGDT to solve a voltage-stability-constrained optimal power flow problem in [16].

To solve a multiyear transmission expansion planning problem under uncertainty of capital costs and load demands, IGDT is utilized to model uncertainty in [17] providing a novel tractable mixed integer linear programming model. In a multiperiod microgrid expansion planning problem, IGDT is used in [18] to model nonrandom uncertainty. The application of IGDT is reported in [19] to model the output power of a wind turbine and outages of the transmission line to obtain a robust design of wide-area power system stabilizers [20–22].

Finally, in [23] short-term scheduling of solar-driven industrial continuous heat treatment furnace is presented using IGDT to model the heat demand uncertainty.

5.2 IGDT Background

Uncertainty is one of the inevitable problems in power systems that may occur for different parameters such as pool price and consumer demand [24]. As discussed earlier, uncertainty may have undesirable or desirable effects. IGDT develops two robustness and opportunity functions to evaluate the undesired and desired aspects of uncertainty, respectively. By considering the uncertainty parameter α and a set of decision variables Q, opportunity and robustness functions $\widehat{\beta}$ and α can be considered as (5.1) and (5.2), respectively:

$$\widehat{\beta} = \min_{\alpha} \{\alpha : \text{sweeping success is sometimes enabled}\}, \tag{5.1}$$

$$\widehat{\alpha} = \max_{\alpha} \{\alpha : \text{minimal requirements are always satisfied}\}. \tag{5.2}$$

The final decision should be made by the decision maker based on the quantitation obtained from the robustness and opportunity functions. Expressions (5.1) and (5.2) should be modified according to the considered problems, which are presented in the following subsections. IGDT consists of three parts: system model, operation requirements, and uncertainty modeling.

5.2.1 *System Modeling*

The input/output structure of a system is presented in system model $R(q, \lambda)$ to apply the decision that is made. This expression is used to assess the response of the system, considering the choices of the decision maker q and the uncertainty parameter λ. In the power procurement problem of a large consumer, which is studied in this chapter, the power price in the pool market is the uncertainty parameter λ, the amount of procured power from the alternative sources are the decision variables, and power procurement cost function is treated as the system model.

5.2.2 *Operation Requirements*

The anticipations or requirements of the system that can be treated as costs are described by performance requirements. The opportunity and robustness functions are used to evaluate the requirements as shown by expressions (5.1) and (5.2). Considering the power procurement problem of a large consumer, the opportunity and robustness functions of the problem can be modified as follows:

$$\widehat{\beta} = \min_{\alpha} \left\{ \begin{array}{l} \alpha : \text{minimum cost of power procurement is less than} \\ \text{a given cost level} \end{array} \right\}, \tag{5.3}$$

$$\overset{\frown}{\alpha} = \max_{\alpha} \left\{ \begin{array}{l} \alpha : \text{maximum cost of power procurement is no} \\ \text{higher than a given cost target} \end{array} \right\}. \tag{5.4}$$

The opportunity function considers the lowest value of α that gives the minimum cost of power procurement of a large consumer, which could be as small as the deterministic case results. This function is used to obtain risk-taking strategies. It should be noted that the best model anticipations might be very pessimistic under severe uncertainty operating conditions. Equation (5.3) can be rewritten as follows:

$$\widehat{\beta}(r_w) = \min_{\alpha} \left\{ \alpha : \min_{\alpha} \left(R(q, \lambda) \right) \leq r_w \right\}. \tag{5.5}$$

It should be noted that the small amount of $\widehat{\beta}(r_w)$ is favorable, which represents a possible benefit even without low prices.

The greatest level of uncertainty is expressed by the robustness function, in which the cost of power procurement cannot be greater that a given amount. Therefore, Eq. (5.6) can be defined mathematically as follows:

$$\overset{\frown}{\alpha}(r_C) = \max_{\alpha} \left\{ \alpha : \max_{\alpha} \left(R(q, \lambda) \right) \leq \mathrm{C}_l \right\}. \tag{5.6}$$

A higher $\alpha(\overset{\frown}{r_C})$ means that the decision that is made is risk-averse, robust, and insensitive to uncertainty, while smaller$\alpha(\overset{\frown}{r_C})$ denotes fragile decisions.

5.2.2.1 Uncertainty Modeling

Based on prior information about the uncertainty vector λ, an IGDT model can be used to model uncertainty. As stated previously, a fractional error model is used to analyze the impact of power price uncertainty in a market. The uncertainty model is detailed in the following sections.

5.3 Mathematical Formulation

In this section, the IGDT method is applied to solve the power procurement problem faced by a large consumer considering power price uncertainty in the pool market model.

5.3.1 Decision Variables

As stated earlier, alternative sources of power procurement (bilateral contracts, pool market, photovoltaic systems, wind turbines, self-generating facilities) are the decision variables:

$$q_t = \left[P_t^p, P_{l,t}^{\mathrm{BC}}, P_{j,h,t}, P_t^{\mathrm{charge}}, P_t^{\mathrm{disc}}, P_t^{\mathrm{wind}}, P_t^{\mathrm{PV}}\right]; t = 1, \ldots, 24 \tag{5.7}$$

where P_t^p is energy procured from pool market, $P_{l,t}^{\mathrm{BC}}$ is power procured from bilateral contracts, $P_{j,h,t}$ is power produced by jth unit of self-generating units, $P_t^{\mathrm{charge}}/P_t^{\mathrm{disc}}$ is the charging/discharging power of battery storage, P_t^{wind} is power produced from wind turbines, P_t^{PV} is power produced by PV system.

5.3.2 Model of Procurement Cost

The cost function of the electricity procurement problem faced by a large consumer is provided by Eq. (5.8):

$$R(q, \lambda) = \sum_{l=1}^{B}\sum_{t=1}^{T}\lambda_{l,t}P_{l,t}^{\mathrm{BC}} + \sum_{t=1}^{T}\left\{\begin{array}{l} \lambda_t P_t^p + \sum_{j=1}^{N_j}\sum_{h=1}^{N_h} S_{j,h}^{\mathrm{DG}} P_{j,h,t} \\ +C_t^B \times \left(\chi \times P_t^{\mathrm{charge}} + \dfrac{P_t^{\mathrm{disc}}}{\eta}\right) \\ +C_t^{\mathrm{wind}} \times P_t^{\mathrm{wind}} + C_t^{\mathrm{PV}} \times P_t^{\mathrm{PV}} \end{array}\right\}. \tag{5.8}$$

The model presented in Eq. (5.8) is detailed in Chap. 2.

5.3.3 Uncertainty Modeling

As stated earlier, the power price in the pool market is treated as an uncertainty parameter. In this case, the fractional error model of the info-gap method is expressed as Eq. (5.9) to model the uncertainty of the market price:

$$U\left(\alpha, \widetilde{\lambda}_t\right) = \left\{\lambda_t : \frac{\left|\lambda_t - \widetilde{\lambda}_t\right|}{\widetilde{\lambda}_t} \leq \alpha\right\}, \alpha \geq 0, \tag{5.9}$$

where $\widetilde{\lambda}_t$ is the estimation of the pool price at time t (\$/MWh), and α is the uncertainty parameter modeling the gap size between the unknown and the known. According to Eq. (5.9), the fractional deviation of the power prices from estimated values $\widetilde{\lambda}_t$ cannot be greater than α.

5.3.4 Robustness Function

In the robustness function, the maximum value of resistance to power price uncertainty is shown by $\widehat{\alpha}(C_r)$. The large consumer with robust performance, in paying more money, wants to choose a risk-averse strategy [5]. It should be noted that the large value of $\widehat{\alpha}(C_r)$ is desirable. In solving the optimization problem, $\widehat{\alpha}(C_r)$ can be calculated using Eq. (5.10):

$$\widehat{\alpha}(r_c) = \max_{q} \alpha(r_c, q). \tag{5.10}$$

To compute the robustness function, the maximum power procurement cost should be determined. Based on Eq. (5.9), high power prices in the market can be expressed as

$$\lambda_t = \widetilde{\lambda}_t + \alpha\widetilde{\lambda}_t. \tag{5.11}$$

The robustness function can be calculated by substituting (5.11) into (5.8) and solving for α, which is shown by (5.12):

$$\max\left\{\sum_{l=1}^{B}\sum_{t=1}^{T}\lambda_{l,t}P_{l,t}^{\mathrm{BC}} + \sum_{t=1}^{T}\left\{\begin{array}{l} (\widetilde{\lambda}_t + \alpha\widetilde{\lambda}_t)P_t^{p} + \sum_{j=1}^{N_j}\sum_{h=1}^{N_h} S_{j,h}^{\mathrm{DG}} P_{j,h,t} \\ + C_t^{B} \times \left(\chi \times P_t^{\mathrm{charge}} + \dfrac{P_t^{\mathrm{disc}}}{\eta}\right) \\ + C_t^{\mathrm{wind}} \times P_t^{\mathrm{wind}} + C_t^{\mathrm{PV}} \times P_t^{\mathrm{PV}} \end{array}\right\}\right\} \le r_c, \tag{5.12}$$

$$\begin{array}{l}\text{Subject to:} \\ \text{constraints } (2.2) - (2.27).\end{array} \tag{5.13}$$

To consider critical power procurement costs that are higher than the minimum procurement cost (deterministic case), the function $\widehat{\alpha}(r_c)$ can be considered. It should be noted that this function is used to arrive at risk-averse strategies. Therefore, to obtain a higher degree of robustness against power price uncertainty, higher values for the robustness function should be considered.

5.3.5 Opportunity Function

The opportunity function is used to develop risk-taker strategies. In other words, to determine the possible reduction of power procurement costs because of low power price in the market, the opportunity function can be used. In this way, small values of $\widehat{\beta}(r_w)$ are desirable. Equation (5.14) presents the mathematical formulation of the opportunity function:

$$\widehat{\beta}(r_w) = \min_q \alpha(r_w, q), \tag{5.14}$$

where r_w is the cost level for the opportunity function ($). It should be noted that r_w is smaller than r_c. The power price reduction can be modeled using the following equation, which is similar to the consideration in the robustness function:

$$\lambda_t = \widetilde{\lambda}_t - \alpha\widetilde{\lambda}_t. \tag{5.15}$$

The opportunity strategy can be calculated by substituting (5.15) into (5.8) and solving for α, which is shown by (5.16):

$$\min\left\{\sum_{l=1}^{B}\sum_{t=1}^{T}\lambda_{l,t}P_{l,t}^{\text{BC}} + \sum_{t=1}^{T}\left\{\begin{array}{l}(\widetilde{\lambda}_t - \alpha\widetilde{\lambda}_t)P_t^p + \sum_{j=1}^{N_j}\sum_{h=1}^{N_h} S_{j,h}^{\text{DG}} P_{j,h,t} \\ +C_t^B \times \left(\chi \times P_t^{\text{charge}} + \dfrac{P_t^{\text{disc}}}{\eta}\right) \\ +C_t^{\text{wind}} \times P_t^{\text{wind}} + C_t^{\text{PV}} \times P_t^{\text{PV}}\end{array}\right\}\right\} \leq r_w, \tag{5.16}$$

Subject to:
constraints (2.2) – (2.27). (5.17)

It should be noted that small values of $\widehat{\beta}(r_w)$ mean that the corresponding decision has low resistance to experiencing low power procurement cost levels. Therefore, a windfall benefit can be attained.

Finally, the large consumer can assess the costs as being near minimum costs via opportunity and robustness using the IGDT method. The problem is formulated as a mixed-integer nonlinear programming problem and solved using a GAMS optimization program.

5.4 Numerical Simulation

To solve the power procurement problem faced by a large consumer using the IGDT method, the case study considered in Chap. 2 is used. All the required information about the case study can be found in Chap. 2.

5.4.1 *Simulation Results*

As stated earlier, the power procurement problem of a large consumer is solved under power price uncertainty in the market using the IGDT method. The results obtained are presented using three different strategies: risk-averse, risk-neutral, and risk-taker corresponding to robust, deterministic, and opportunity functions, respectively, each of which is solved in two cases respectively without and with consideration of the impact of demand response programs. It should be noted that time-of-use (TOU) demand response rate (DRP), which is detailed in previous chapters, is used to reduce power procurement costs.

5.4.2 *Risk-Neutral Results*

Equations (2.1)–(2.27) present the formulation of the risk-neutral strategy. In this strategy, which presents deterministic case results, the power procurement result is equal to $40,253.252 and $38,401.038 respectively without and with implementation of demand response programs. A summary of the obtained results is provided in Table 5.1, which is detailed in Chap. 2. It is obvious that by implementing the TOU-DRP, total power procurement costs are reduced by about 4.6%.

5.4.3 *Risk-Averse Results*

As stated earlier, the robustness function in the IGDT method is used to model the pernicious effects of uncertainty and derive a risk-averse strategy. The robust performance of the large consumer is obtained solving (5.10) subject to (5.11)–(5.13) under power price uncertainty in the market. Robustness procurement cost

Table 5.1 Results of risk neutral strategy

Different costs	Without DRP	With DRP
Procurement cost from pool market ($)	13,168.363	11,350.531
Procurement cost from BCs ($)	9,322.080	9,322.080
Procurement cost from SG units ($)	17,762.808	17,728.427
Total procurement cost ($)	40,253.252	38,401.038

against the robustness function $\widehat{\alpha}(r_c)$ (in percentage terms) is presented in Fig. 5.1 without and with consideration of the DRP, respectively.

Results obtained in the robustness function can be used by the large consumer to adopt the risk-averse strategy. From Fig. 5.1 it is obvious that by increasing the robustness cost r_c, robustness is increased, as expected. In addition, it can be inferred to get the higher robustness against the power price uncertainty, the higher cost should be paid. Conversely, more cost payment by consumer will result in a more robust strategy becomes. In the first case, without DRP, by paying $42,000 to procure electricity, 12.16% robustness against power price uncertainty can be achieved. Also, in the second case, assuming $40,000 as the procurement cost, 11.4% robustness is achieved against power price increases in the pool market.

Finally, it should be noted that the results obtained in this section could be used as a risk-averse strategy.

5.4.4 *Risk-Taker Results*

To investigate the positive effects of uncertainty and present the risk-taker strategy, the opportunity function is developed in the IGDT method. Figure 5.2 shows the opportunity function results considering a reduction in the power price in the pool market. Figure 5.2, makes it clear that using DRP has reduced the procurement costs

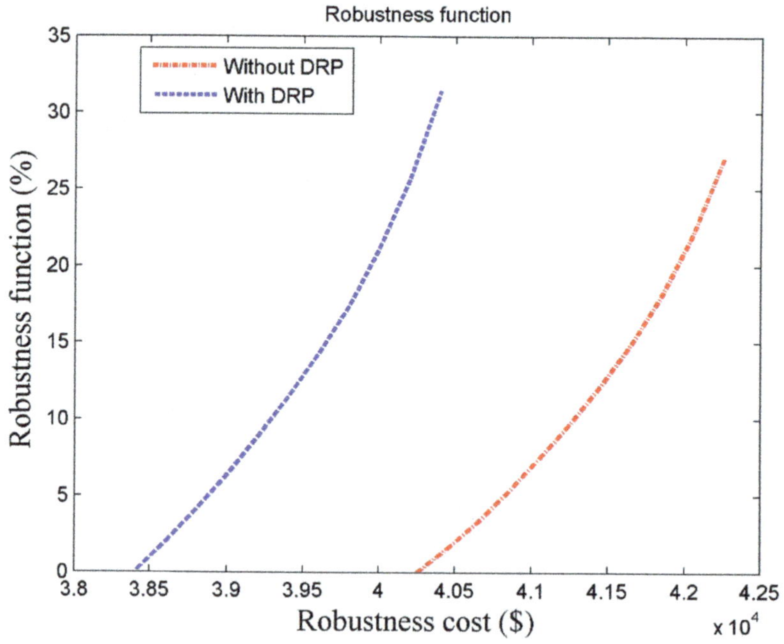

Fig. 5.1 Robustness cost for without and with considering DRP

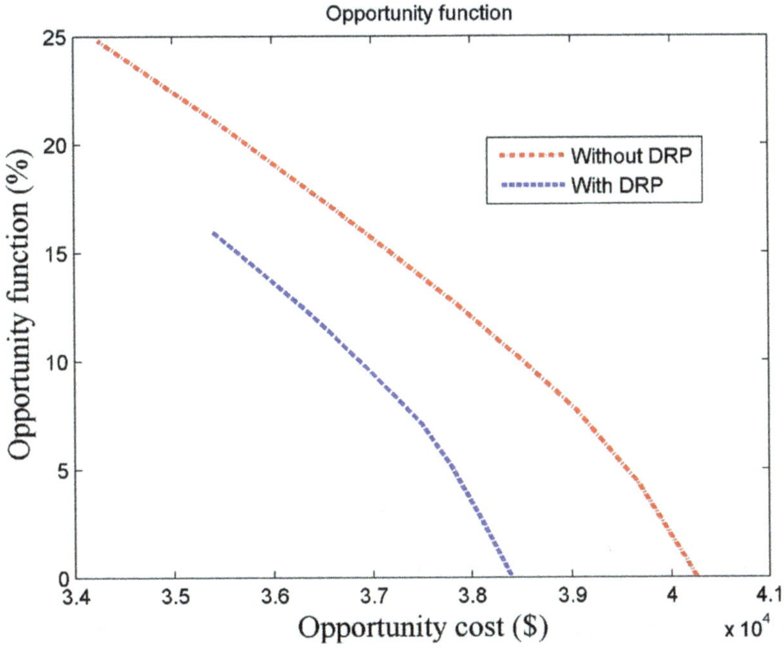

Fig. 5.2 Opportunity cost for without and with considering DRP

for a lower pool market price. For example, for a 15% power price reduction in the market, consumer costs will be \$37,250 and \$35,700 respectively without and with DRP cases, which means implementing DRP has reduced procurement costs by about 4.2%.

5.4.5 *Analyzing Alternative Sources of Power Procurement for Large Consumer*

The impact of power price uncertainty on the procurement of power from alternative sources is investigated in the following subsection using three strategies: risk-taker, risk-neutral, and risk-averse.

5.4.5.1 Load Profile

In the three strategies, the load profile has changed as a result of implementing DRP and considering a market price change. Figure 5.3 shows the new load profile in the three strategies. According to this figure, the load is shifted from peak periods to

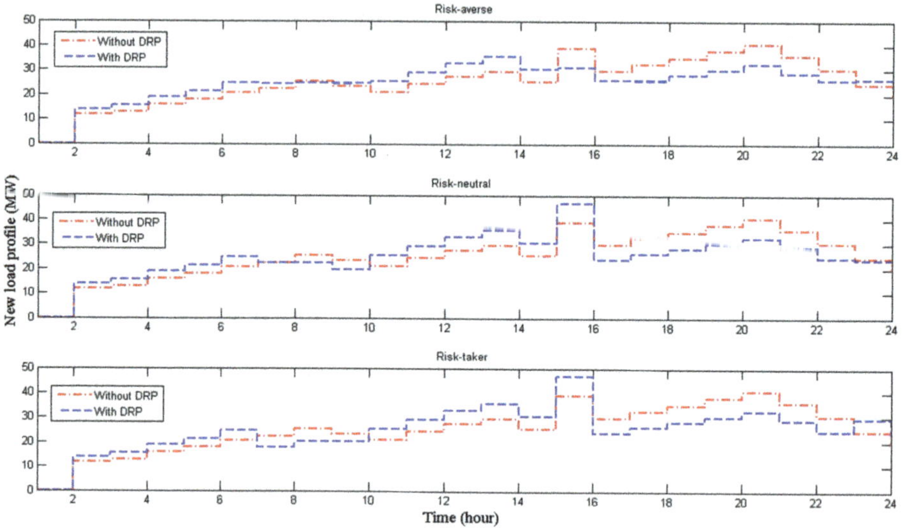

Fig. 5.3 New load profile in the three strategies

off-peak periods, which occurred as a result of implementing TOU-DRP. In addition, in the risk-averse strategy, the load curve is flatter in comparison with the other strategies, especially at the peak periods.

5.4.5.2 Power Purchased in Pool Market

Figure 5.4 illustrates the power purchased in the pool market. As can be seen in the figure, there is a significant reduction in power purchased in the pool market. It should be noted that the risk-averse strategy investigates the impact of pool price increases. In addition, in the risk-taker strategy, which considers a reduction in power price, the power purchased in the pool market is higher than the risk-neutral strategy. Finally, by implementing DRP, at peak periods power purchased in the pool market is reduced.

5.4.5.3 Power Purchased Under Bilateral Contracts

Bilateral contracts are treated as one of the power procurement sources to meet the required demand of the large consumer. Figure 5.5 presents the power purchased under bilateral contracts for the three strategies. The results show that in the risk-averse strategy, more power is procured in comparison with the other strategies. In addition, the procured power in the risk-taker strategy is equal to zero because required demand is supplied by power purchased in the pool market in the low-power-price situation.

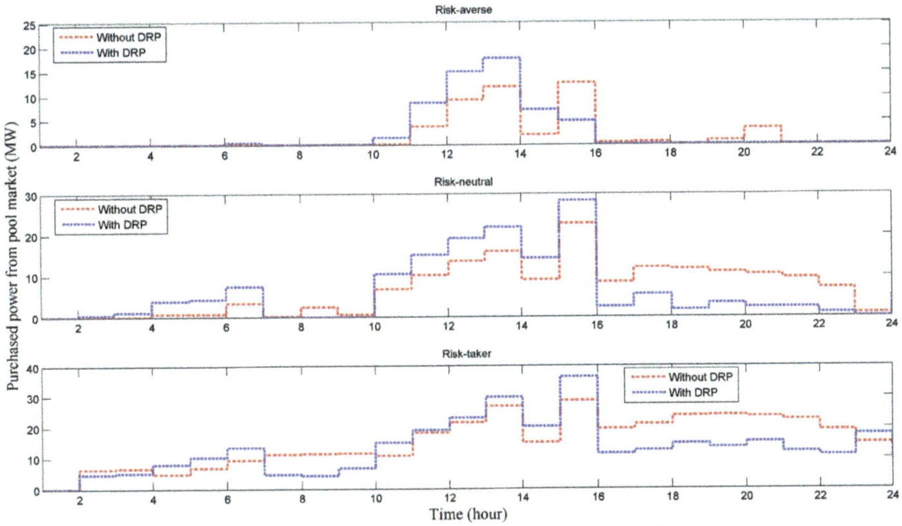

Fig. 5.4 Purchased power from the pool market in the three strategies

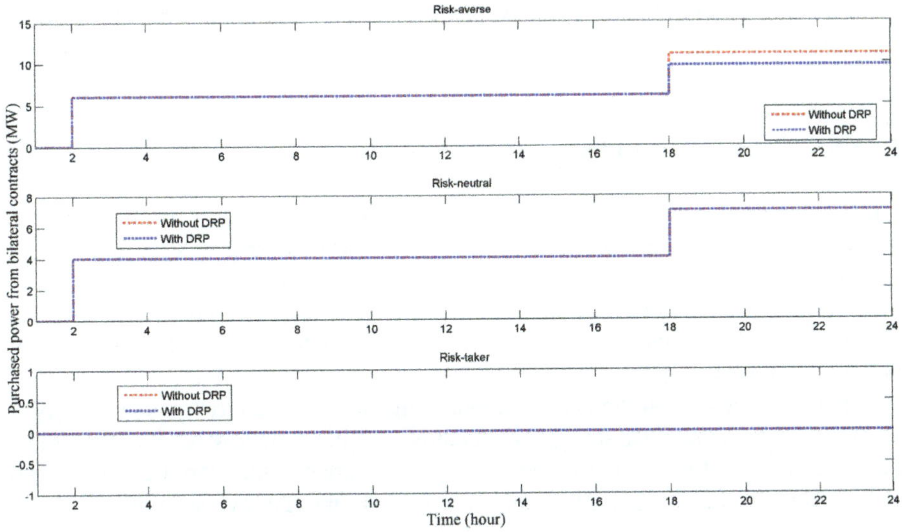

Fig. 5.5 Procured power from the bilateral contracts units in the three strategies

5.4.5.4 Power Purchased from Self-Generating Units

In the considered case study, some part of the required load demand of the large consumer is procured from self-generating units. The power purchased from self-generating units is depicted in Fig. 5.6. In the risk-averse strategy, owing to power price increases in the pool market, more power is purchased from these units.

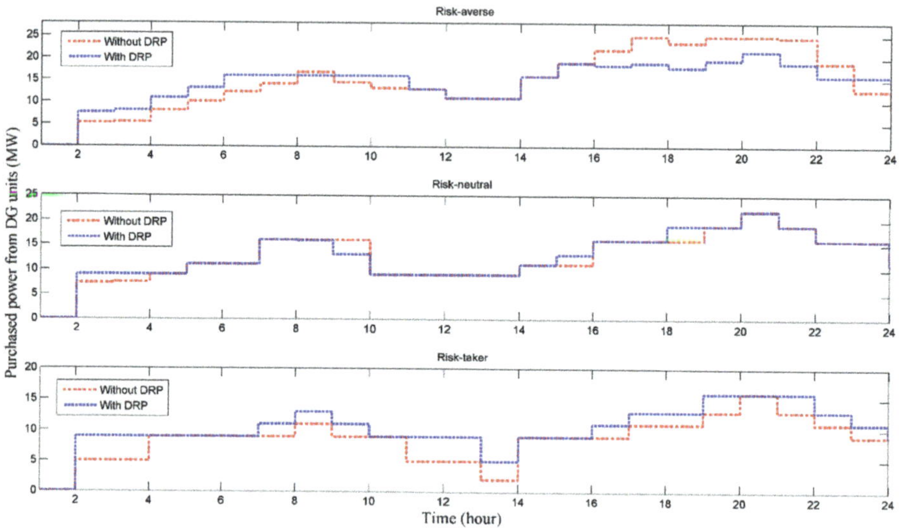

Fig. 5.6 Procured power from the self-generating units

Also, the impact of the DRP on the power from purchased microturbines is quite moderate.

5.4.5.5 Power Procured from Storage Systems

Finally, battery storage is utilized to store energy during off-peak periods and use in peak periods. Figures 5.7 and 5.8 illustrate the charge/discharge state and energy stored in batteries for the three strategies. The results obtained show that in the risk-averse strategy, the number and time of battery storage charge/discharge are less than in the other two strategies, while these amounts are greater in the risk-taker strategy than in the other two.

Figure 5.8 shows that the energy stored in the battery in the risk-taker strategy is more than that in the other strategies, which is a result of the low pool market price. In addition, it is obvious that implementing DRP reduced the stored energy of the battery in all of the considered strategies, especially the risk-averse strategy.

5.5 Conclusion

To solve the power procurement problem of a large consumer under power price uncertainty, the IGDT method is implemented to model the positive and negative impacts of uncertainty. To do so, based on the opportunity and robustness function of the IGDT method, three different strategies, risk-averse, risk-neutral, and risk-taker,

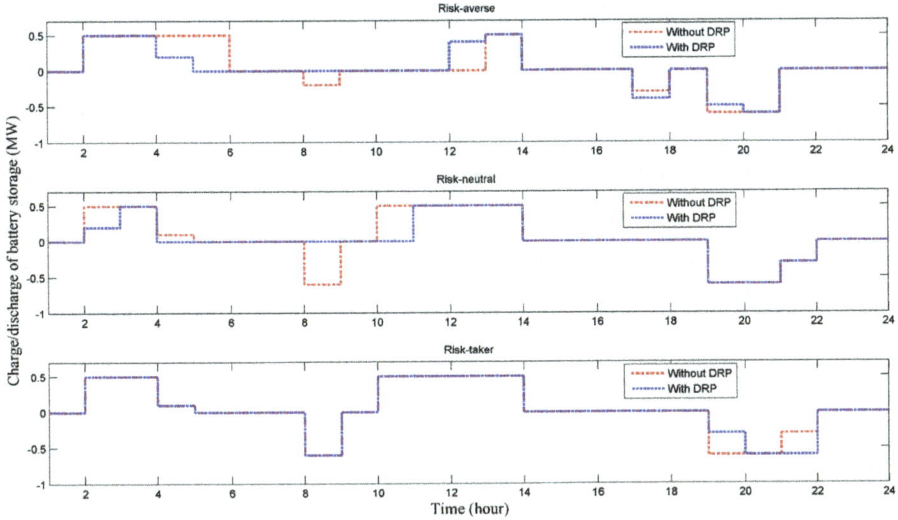

Fig. 5.7 Charging/discharging of the battery in the three strategies

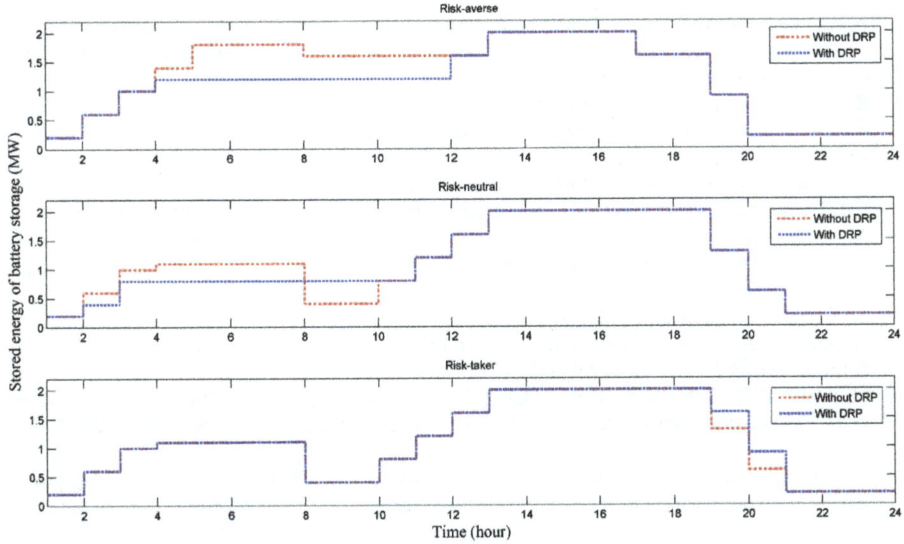

Fig. 5.8 Stored energy in the battery in the three strategies

are formulated to assess the risk of the large consumer's power procurement process. The considered method is that implemented by the sample large consumer model introduced in Chap. 2.

According to the obtained results, in the risk-neutral strategy, the total cost of power procured from alternative sources was \$40,253 and \$38,401 without and with consideration of DRP, respectively, which means by implementing TOU-DRP, total power procurement costs reduced by about 4.6%.

In the risk-averse strategy, the large consumer is 12.16% robust against power price uncertainty without implementing DRP by paying $42,000 to procure electricity. In addition, with $40,000 as the procurement cost, 11.4% robustness is achieved against the power price increase in the pool market, implementing the TOU-DRP.

In the risk-taker strategy, for a 15% power price reduction, the large consumer's costs will be $37,250 and $35,700, respectively without and with consideration of DRP cases, which means implementing DRP reduced procurement costs by about 4.2%.

References

1. Y. Ben-Haim, Y. Ben-Haim, *Info-Gap Decision Theory: Decisions Under Severe Uncertainty* (Academic Press, New York, 2006)
2. A.J. Conejo, M. Carrión, J.M. Morales, *Decision Making Under Uncertainty in Electricity Markets*, vol 153 (Springer, Boston, 2010)
3. I. Ahmadian, O. Abedinia, N. Ghadimi, Fuzzy stochastic long-term model with consideration of uncertainties for deployment of distributed energy resources using interactive honey bee mating optimization. Front. Energy **8**(4), 412–425 (2014)
4. Z. Chen, L. Wu, Y. Fu, Real-time price-based demand response management for residential appliances via stochastic optimization and robust optimization. IEEE Trans. Smart Grid **3**(4), 1822–1831 (2012)
5. S. Nojavan, H. Ghesmati, K. Zare, Robust optimal offering strategy of large consumer using IGDT considering demand response programs. Electr. Power Syst. Res. **130**, 46–58 (2016)
6. A. Najafi-Ghalelou, S. Nojavan, K. Zare, Heating and power hub models for robust performance of smart building using information gap decision theory. Int. J. Electr. Power Energy Syst. **98**, 23–35 (2018)
7. S. Nojavan, K. Zare, M.A. Ashpazi, A hybrid approach based on IGDT–MPSO method for optimal bidding strategy of price-taker generation station in day-ahead electricity market. Int. J. Electr. Power Energy Syst. **69**, 335–343 (2015)
8. S. Nojavan, K. Zare, M.R. Feyzi, Optimal bidding strategy of generation station in power market using information gap decision theory (IGDT). Electr. Power Syst. Res. **96**, 56–63 (2013)
9. S. Nojavan, K. Zare, Risk-based optimal bidding strategy of generation company in day-ahead electricity market using information gap decision theory. Int. J. Electr. Power Energy Syst. **48**, 83–92 (2013)
10. K. Zare, M.P. Moghaddam, M.K. Sheikh El Eslami, Electricity procurement for large consumers based on information gap decision theory. Energy Policy **38**(1), 234–242 (2010)
11. K. Zare, M.P. Moghaddam, M.K. Sheikh El Eslami, Demand bidding construction for a large consumer through a hybrid IGDT-probability methodology. Energy **35**(7), 2999–3007 (2010)
12. M. Alipour, K. Zare, B. Mohammadi-Ivatloo, Optimal risk-constrained participation of industrial cogeneration systems in the day-ahead energy markets. Renew. Sust. Energ. Rev. **60**, 421–432 (2016)
13. S. Shafiee, H. Zareipour, A.M. Knight, N. Amjady, B. Mohammadi-Ivatloo, Risk-constrained bidding and offering strategy for a merchant compressed air energy storage plant. IEEE Trans. Power Syst. **32**, 1–1 (2016)

14. A. Najafi-Ghalelou, S. Nojavan, K. Zare, Information gap decision theory-based risk-constrained scheduling of smart home energy consumption in the presence of solar thermal storage system. Sol. Energy **163**, 271–287 (2018)
15. J. Zhao, C. Wan, Z. Xu, J. Wang, Risk-based day-ahead scheduling of electric vehicle aggregator using information gap decision theory. IEEE Trans. Smart Grid **8**(4), 1609–1618 (2017)
16. A. Rabiee, S. Nikkhah, A. Soroudi, E. Hooshmand, Information gap decision theory for voltage stability constrained OPF considering the uncertainty of multiple wind farms. IET Renew. Power Gen. **11**(5), 585–592 (2017)
17. A. Kazemi, S. Dehghan, N. Amjady, Multi-objective robust transmission expansion planning using information-gap decision theory and augmented ⍰-constraint method. IET Gener. Transm. Distrib. **8**(5), 828–840 (2014)
18. X. Cao, J. Wang, B. Zeng, A chance constrained information-gap decision model for multi-period microgrid planning. IEEE Trans. Power Syst. **33**(3), 2684–2695 (2018)
19. D. Ke, F. Shen, C.Y. Chung, C. Zhang, J. Xu, Y. Sun, Application of information gap decision theory to the design of robust wide-area power system stabilizers considering uncertainties of wind power. IEEE Trans. Sustain. Energy **9**(2), 805–817 (2018)
20. M. Eslami, H.A. Moghadam, H. Zayandehroodi, N. Ghadimi, A new formulation to reduce the number of variables and constraints to expedite scuc in bulky power systems. Proc. Natl. Acad. Sci. India Sect. A: Phys. Sci., 1–11 (2018). https://doi.org/10.1007/s40010-017-0475-1
21. O. Abedinia, M. Bekravi, N. Ghadimi, Intelligent controller based wide-area control in power system. Int. J. Uncertainty Fuzziness Knowl. Based Syst. **25**(01), 1–30 (2017)
22. A. Jalili, N. Ghadimi, Hybrid harmony search algorithm and fuzzy mechanism for solving congestion management problem in an electricity market. Complexity **21**(S1), 90–98 (2016)
23. F. Jabari, S. Nojavan, B. Mohammadi-ivatloo, H. Ghaebi, H. Mehrjerdi, Risk-constrained scheduling of solar Stirling engine based industrial continuous heat treatment furnace. Appl. Therm. Eng. **128**, 940–955 (2018)
24. M. Charwand, Z. Moshavash, Midterm decision-making framework for an electricity retailer based on information gap decision theory. Int. J. Electr. Power Energy Syst. **63**, 185–195 (2014)

Chapter 6
Energy Procurement via Hybrid Robust-Stochastic Approach

Raouf Morsali Asl and Ehsan Zargin

Abstract High computational burden is the main disadvantage of the stochastic programming. Also, obtained results are not guaranteed the global optimal solution. On the other hand, by using robust optimization method, only one uncertain parameter can be modeled. In this chapter, a novel hybrid robust-stochastic approach is introduced to benefit advantages of the robust and stochastic methods and overcome the abovementioned problems. The application of the robust-stochastic approach is investigated in the power procurement problem of a large consumer, considering uncertainty of load demand, power price, and power output of renewable energy sources as wind turbine and photovoltaic system. The uncertainty load demand and generation of renewable energy sources are modeled by a set of discrete scenarios, while the robust optimization method is used to model power price uncertainty in the pool market. Using considered case study in the previous chapter, the power procurement problem is solved under severe uncertainty, and the numerical analysis is presented.

6.1 Introduction

Applications of stochastic programming and robust optimization on the power procurement problem of a large consumer are well studied in Chaps. 3 and 4, respectively. As discussed in Chap. 3, the stochastic programming has different disadvantages. For example, in order to model uncertainty, detailed information of the uncertainty parameter or parameters is required to generate the scenarios, while enough historical data may not be available for many real-life applications [1, 2]. The other disadvantage of stochastic programming comes from a high number of scenarios when an accurate approximation of distribution is required [3, 4]. It should be noted that a small number of scenarios might lead to inaccurate solutions. In this case, the higher the required number of scenarios, the harder the problem is solved to get an optimal solution [5, 6]. Conversely, a limited number of scenarios

R. M. Asl (✉) · E. Zargin
Young Researchers and Elite Club, Ardabil Branch, Islamic Azad University, Ardabil, Iran

S. Nojavan et al. (eds.), *Robust Energy Procurement of Large Electricity Consumers*,
https://doi.org/10.1007/978-3-030-03229-6_6

may result in an infeasible solution for some realizations of the uncertain parameters. Finally, the optimal global solution may not be attained by using the stochastic programming. The stochastic programming method is detailed in Chap. 3.

To overcome these problems, the robust optimization program can be considered as an alternative method for uncertainty modeling of the parameters that precise probability distribution is difficult to predict [7, 8]. The main advantages of robust optimization methodology can be listed as computational tractability and using rough historical data and experiences of the decision makers to get the boundaries of uncertainty sets [9]. On the other hand, it should be noted that in single-stage uncertainty modeling, only one uncertain parameter can be modeled using the robust optimization method [10, 11]. More information about the robust optimization method can be found in Chap. 4.

As it can be seen in Fig. 6.1, in this chapter, the uncertainty of load demand, wind speed, and solar irradiation are modeled generating scenarios and modeling stochastic programming method. In addition, the power price uncertainty in the pool market is modeled using robust optimization method.

By considering the battery storage system, bilateral contracts with predetermined prices, and micro-turbine as self-generating unit, the power procurement problem faced by a large consumer is solved using developed robust-stochastic approach.

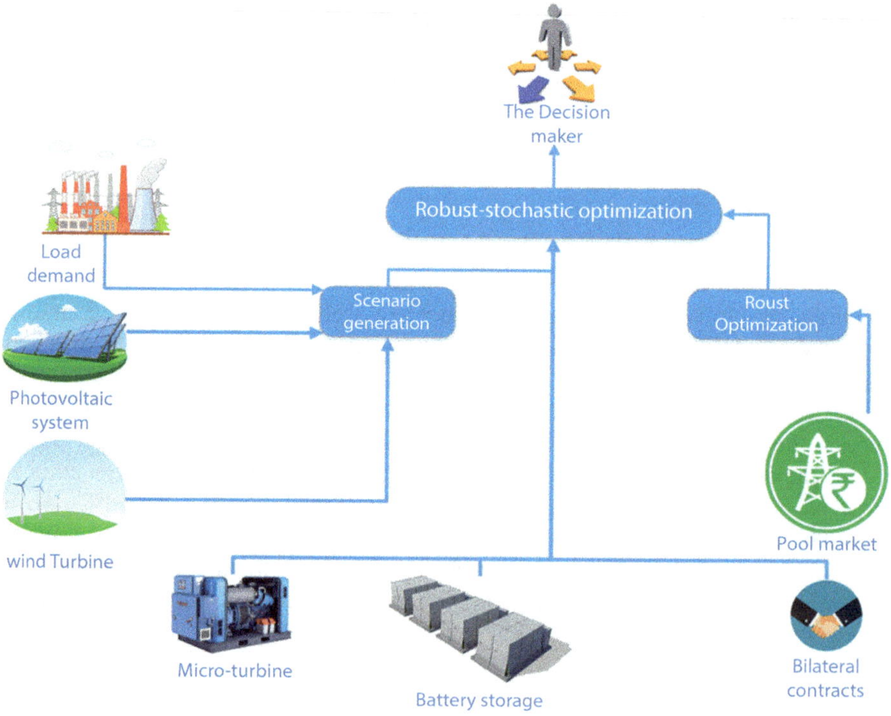

Fig. 6.1 Robust-stochastic structure for power procurement problem of the large consumer

Note that the Weibull distribution is used to generate the wind speed scenarios modeling power output of wind turbine, and the normal distribution is used to generate scenarios of load demand and solar irradiation.

6.2 Robust-Stochastic Approach in the Literature

The introduced robust-stochastic approach is used to solve various problems in different fields of science. In this section, the most worthy and related works to the subject have been reviewed to show the importance and application of the method.

In [12], the hybrid robust-stochastic approach is used to get the optimal demand bidding curves under uncertain information of market and distribution system and network operational limits. The case study in this Ref. includes renewable energy sources and deferrable and inflexible loads. In order to solve the problem, the authors have used decomposition algorithm to get the optimal bid curves minimizing energy procurement cost. Finally, the advantages of the proposed model are illustrated by providing numerical experiments on an example system and the 33-bus system.

The hybrid robust-stochastic approach is utilized to consider stability analysis of DFIG integration on the power system in [13]. In this Ref., using the sensitivity analysis method, the analytic function relationship between elements and stochastic parameters is derived in the state matrix; then, the robust-stochastic stability criterion is derived by constructing the Lyapunov functional including the stochastic parameter model. The application of the proposed model is verified on the IEEE four-generator two-area system and the New England 10-machine 39-bus system in which the proposed method effectively identifies the small-signal stability of the power system under the joint influence of stochastic parameter and external excitation. In addition, the results show that the method can be used for complex system with multiple generators.

There is great interest in practical scheduling of the electric vehicles' charging demand under the worst-case scenario of the wind power generation because of the forecasting error. In this case, the robust scheduling problem of a micro-grid is pursued in [14]. Therefore, the robust-stochastic approach is used to solve the problem as a robust-stochastic shortest path problem. The objective function in this Ref. developed as a weighted sum of the total charging cost and wind power utilization. In addition, to improve the performance in the worst case, a robust simulation-based policy improvement method is introduced. Finally, the performance of the proposed methodology is numerically demonstrated using real information of electric vehicle and wind system.

Rapid increase in uncontrollable renewable energy sources such as photovoltaic power generations is one of the important difficulties in the power system operation. In addition, it should be noted that reduction of controllable resources presents concerns about the reliability of the power system. In this case, the robust-stochastic approach is used to develop new dynamic load dispatch method against the related uncertainties to the problem in [15].

The robust-stochastic approach is used to consider the multi-retailer competition under demand uncertainty in [16]. In this Ref., the retailer competition is modeling using stochastic programming method under the assumption of product substitutability. In addition, the robust deviation, relative robust, and absolute robust problems are studied under uncertainty of the load demand, respectively, using the robust optimization approach. Finally, through numerical examples, the three optimal solutions are compared with different product substitutability.

In [17], utilization of global sensitivity analysis and harmony search algorithm is investigated for optimal design of air-cooled heat exchangers from the economic viewpoint using the hybrid robust-stochastic approach. To identify the non-influential parameters and examine the effect of the design parameters, the global sensitivity analysis is performed which has reduced the size of the optimization problem.

With the impact of uncertainty of the power generation of wind turbine on operation of the power system under demand response program, the robust-stochastic approach is introduced in [18] to build a stochastic scheduling optimization model. In this Ref., firstly, external environment based on the Brownian motion is used for simulation of wind power. Secondly, the built demand response model is developed based on incentive-based and price-based demand responses. Furthermore, the robust-stochastic approach is introduced to build a model for wind power consumption stochastic optimization. The application of proposed method is validated by analyzing the IEEE 36-node 10-unit system connected with 650 MW wind farms. Obtained results show that the robust-stochastic approach is better to overcome wind power uncertainty.

In Ref. [19], the robust-stochastic approach is developed for energy management of residential micro-grid. Heating and electricity technology and combined cooling are considered to meet various load demands. The optimal installed capacity investment and operation control of CCHP (combined cooling, heating, and power) are determined using the two-stage programming method. In addition, interval robust solutions are obtained under different robustness levels using interval programming and robust-stochastic optimization methods which are feasible for uncertain data. Obtained results in this Ref. can be used to help the operators of micro-grid minimizing the operation and investment cost.

A strategic bidding model of several price-taker plug-in EV aggregators sharing the same distribution network is derived in [20], for both ancillary services (up-/downregulation reserve) and day-ahead energy markets. It should be noted that the strategic feasible space of an aggregator is dependent on performance of other aggregators. The objective function of aggregators is concentrated on cost of purchased energy from the real-time and day-ahead market minus the revenue from the day-ahead regulation market. In this Ref., a hybrid robust-stochastic approach is developed to deal with different uncertainties including day-ahead up-/downregulation prices, real-time energy prices, and day-ahead energy prices faced by an aggregator in the bidding strategy problem. It should be noted that the day-ahead prices are modeled by generating different discrete scenarios, while real-time prices are represented by the confidence bounds. Obtained results of considered case study demonstrate the applicability and tractability of the proposed model.

In order to determine the selling price to end-user consumers, the electricity retailer considered real-time pricing, time-of-use pricing, and fixed pricing in [21]. Furthermore, the offering and bidding curves are prepared to bid and offer to the day-ahead market and provide power exchange between the power market and retailer. To do so, the robust optimization approach (ROA) is utilized in this Ref. to obtain optimal offering and bidding strategies for the retailer. It should be noted that the uncertainty of power market price is modeled using the ROA considering the maximum and minimum limits of prices for uncertainty modeling.

The proposed robust mixed-integer linear programming model can be used to help retailer choose a robust decision in the presence of market price uncertainty. Meanwhile, the stochastic programming is used to model uncertainty of volatile climate condition and customer's demand. To validate the proposed robust optimization model, three case studies are evaluated, and the results are compared.

In Ref. [22], a general model framework based on the robust-stochastic approach is developed to get the optimal offering strategy for concentrating solar power (CSP) plants in regulation, reserve, and joint day-ahead energy markets. The derived strategy is robust against the uncertainty of the solar energy and market price uncertainty which is modeled using the stochastic programming method. The case study results demonstrate the validity of the proposed model.

Finally, in [23], a methodology based on the hybrid robust-stochastic approach is developed to derive the optimal offering curves for a concentrating solar power plant. In this Ref., the uncertainty of the thermal production from the solar field is modeled using robust optimization method, and volatility of market prices is modeled through different scenarios using the stochastic programming method. In this paper, the owner of the solar plant is a price-taker producer participating in a pool-based electricity market to maximize its expected profit. Optimal offering curves are derived developing a mixed-integer linear programming model which is robust against the uncertainty associated with the thermal production of the solar field and uncertain market prices.

In this chapter, in order to avoid the disadvantages of robust optimization and stochastic programming, a novel robust-stochastic programming method is introduced. In the power procurement problem of a large consumer, considering the sample case study in Chap. 2, uncertainties of the load demand and power output of renewable energy sources as photovoltaic (PV) and wind turbine (WT) are modeled using stochastic programming method, and the power price uncertainty is modeled via robust optimization method. It should be noted that the normal distribution is used to generate scenarios of load demand and power output of PV system and the Weibull distribution to model the wind speed scenarios.

6.3 Problem Formulation

The deterministic formulation of power procurement problem is present in (2.1)–(2.27). In this chapter, the problem is solved without considering demand response programs. Therefore, the problem (6.1)–(6.21) represents the deterministic power procurement problem. In the following, the formulation (2.1)–(2.22) is presented:

$$\underset{P_t^p, P_{l,t}^{\mathrm{BC}}, P_{j,h,t}, P_t^{\mathrm{charge}}, P_t^{\mathrm{disc}}, P_t^{\mathrm{wind}}, P_t^{\mathrm{PV}}}{\text{Minimize}} \sum_{l=1}^{B} \sum_{t=1}^{T} \lambda_{l,t} P_{l,t}^{\mathrm{BC}} + \sum_{t=1}^{T} \left\{ \begin{array}{l} \lambda_t P_t^p + \sum_{j=1}^{N_j} \sum_{h=1}^{N_h} S_{j,h}^{\mathrm{DG}} P_{j,h,t} \\ + C_t^B \times \left(\chi \times P_t^{\mathrm{charge}} + \dfrac{P_t^{\mathrm{disc}}}{\eta} \right) \\ + C_t^{\mathrm{wind}} \times P_t^{\mathrm{wind}} + C_t^{\mathrm{PV}} \times P_t^{\mathrm{PV}} \end{array} \right\} \tag{6.1}$$

$$\sum_{l=1}^{B} P_{l,t}^{\mathrm{BC}} + P_t^p + \sum_{j=1}^{N_j} \sum_{h=1}^{N_h} P_{j,h,t} + P_t^{\mathrm{wind}} + P_t^{\mathrm{PV}} + P_t^{\mathrm{disc}} = \mathrm{load}_t^0 + P_t^{\mathrm{charge}} \tag{6.2}$$

$$P_{l,t}^{\min} s_l < P_{l,t}^{\mathrm{BC}} < P_{l,t}^{\max} s_l; \forall l, t \tag{6.3}$$

$$P_t^{\mathrm{BC}} = \sum_{l=1}^{N_l} P_{l,t}^{\mathrm{BC}}; \forall t \tag{6.4}$$

$$\mathrm{Cost} = \sum_{t=1}^{T} \sum_{j=1}^{N_j} \sum_{h=1}^{N_h} S_{j,h}^{\mathrm{DG}} P_{j,h,t} \tag{6.5}$$

$$0 \le P_{j,h,t} \le \left(P_{j,h}^{\mathrm{MAX}} - P_{j,h-1}^{\mathrm{MAX}} \right) \times U_{j,t}; \forall j, h, t \tag{6.6}$$

$$0 \le P_{j,1,t} \le P_{j,1}^{\mathrm{MAX}} \times U_{j,t}; \forall j, t \tag{6.7}$$

$$\sum_{h=1}^{N_h} P_{j,h,t} - \sum_{h=1}^{N_h} P_{j,h,t-1} \le R_j^{\mathrm{up}} \times U_{j,t}; \forall j, t \tag{6.8}$$

$$\sum_{h=1}^{N_h} P_{j,h,t-1} - \sum_{h=1}^{N_h} P_{j,h,t} \le R_j^{\mathrm{down}} \times U_{j,t-1}; \forall j, t \tag{6.9}$$

$$\begin{array}{l} U_{j,t} - U_{j,t-1} \le U_{j,t+\mathrm{Up}_{j,i}}; \forall j, \forall t, \forall i \\ U_{j,t-1} - \mathrm{U}_{j,t} \le 1 - \mathrm{U}_{j,t+\mathrm{Dn}_{j,i}}; \forall j, \forall t, \forall i \end{array} \tag{6.10}$$

$$\begin{aligned} \mathrm{Up}_{j,i} &= \begin{Bmatrix} i & i \le \mathrm{MUT}_j \\ 0 & i \succ \mathrm{MUT}_j \end{Bmatrix} \\ \mathrm{Dn}_{j,i} &= \begin{Bmatrix} i & i \le \mathrm{MDT}_j \\ 0 & i \succ \mathrm{MDT}_j \end{Bmatrix} \end{aligned} \tag{6.11}$$

$$P_{t,s}^{\mathrm{wind,max}} = \begin{cases} 0 & V_{t,s}^{w} < V_{\mathrm{ci}} \\ p_r \times \left(\frac{V_{t,s}^{w} - V_{\mathrm{ci}}}{V_r - V_{\mathrm{ci}}}\right)^3 & V_{\mathrm{ci}} < V_{t,s}^{w} < V_{\mathrm{cr}} \\ p_r & V_r < V_{t,s}^{w} < V_{c0} \\ 0 & V_{t,s}^{w} > V_{c0} \end{cases} \tag{6.12}$$

$$P_t^{\mathrm{wind}} \le P_t^{\mathrm{wind,max}}; \forall \mathrm{t} \tag{6.13}$$

$$P_t^{M,\max} = \frac{G_t^a}{G_{a_0}} \times \left\{ P_{\mathrm{Max},0}^{M} + \mu_{P_{\max}} \times \left(T_{t,}^{a} + G_t^a \times \frac{\mathrm{NOCT} - 20}{800} - T_{M,0} \right) \right\} \tag{6.14}$$

$$P_t^{\mathrm{PV}} \le P_t^{M,\max}; \forall \mathrm{t} \tag{6.15}$$

$$X_{t_0}^{b} = X_0^{b} \tag{6.16}$$

$$P_t^{\mathrm{charge}} \le P_{\mathrm{charge}}^{\max} \times U_t^{\mathrm{charge}}; \forall t \tag{6.17}$$

$$P_t^{\mathrm{disc}} \le P_{\mathrm{disc}}^{\max} \times U_t^{\mathrm{disc}}; \forall t \tag{6.18}$$

$$X_b^{\min} \le X_t^b \le X_b^{\max}; \forall t \tag{6.19}$$

$$U_t^{\mathrm{charge}} + U_t^{\mathrm{disc}} \le 1; \forall t \tag{6.20}$$

$$X_t^b = X_{t-1}^b + \chi \times P_t^{\mathrm{charge}} - \frac{P_t^{\mathrm{disc}}}{\eta}; \forall t \tag{6.21}$$

In the abovementioned formulation, Eq. (6.1) presents the total power procurement cost as the objective function. Constraint (6.2) provides the power balance constraint between required demand and procured power from alternative sources. The operational requirements of the problem are presented using constraints (6.3)–(6.21).

In the model (6.1)–(6.21), $P_{l,t}^{\mathrm{BC}}$ is the procured power from the BCs with predetermined price $\lambda_{l,t}$; P_t^p is the cost of energy procurement from the pool market with the market price λ_t; $P_{j,h,t}$ is the cost of produced power by the jth unit of the self-generation units; C_t^B is the operating cost of using battery storage, with charging/discharging efficiency χ/η and charging/discharging power of battery storage (MW) $P_t^{\mathrm{charge}}/P_t^{\mathrm{disc}}$; C_t^{wind} is the cost of produced power from the wind turbine; and C_t^{PV} is the cost of produced power P_t^{PV} from the PV system.

Also, load_t^0 is the initial load, $P_{l,t}^{\min}/P_{l,t}^{\max}$ is the minimum/maximum capacity relating, P_t^{BC} is the total procured power from the bilateral contracts, $P_{j,h}^{\mathrm{MAX}}$ is the power output of block h of jth unit of self-generating units, R_j^{up}, R_j^{down} is the ramp-up/ramp-down rate limit of j unit, MUT_j,MDT_j is the minimum up-/downtime of self-generation units j, V_t^w is the wind speed, V_r is the rated speed, V_{ci} is the cut-in speed, V_{c0} is the cutout wind speed, $P_t^{M,\max}$ is the maximum power at any temperature T_t^a, G_t^a is irradiation, $P_{\text{Max},0}^M$ is the maximum power, and NOCT is the normal operating cell. For more information about the formulation (6.1)–(6.21), please refer to Chap. 2.

6.4 Hybrid Robust-Stochastic Approach

In this section, the robust-stochastic approach is introduced. It should be noted that power price uncertainty is modeled using robust optimization, and other uncertainties such as load demand and power output of renewable energy sources are modeled using stochastic programming. The robust optimization and stochastic programming methods are detailed in Chaps. 3 and 4, respectively. In the following subsections, the robust-stochastic formulation of the problem is developed.

6.4.1 Robust Formulation

The robust optimization method is detailed in Chap. 4. In the robust formulation of the problem, the only uncertain parameter, power price in the pool market, is located in the objective function (6.1). Therefore, constraints (6.2)–(6.21) remain unchanged. Considering the robust optimization (4.5)–(4.13), the robust model in the presence of the uncertainty is provided as follows:

$$\text{Min OF} = z \cdot \Gamma + \sum_{t=1}^{24} q_t + \sum_{l=1}^{B} \sum_{t=1}^{T} \lambda_{l,t} P_{l,t}^{\text{BC}} + \sum_{t=1}^{T} \left\{ \begin{array}{l} \lambda_t P_t^p + \sum_{j=1}^{N_j} \sum_{h=1}^{N_h} S_{j,h}^{\text{DG}} P_{j,h,t} \\ + C_t^B \times \left(\chi \times P_t^{\text{charge}} + \dfrac{P_t^{\text{disc}}}{\eta} \right) \\ + C_t^{\text{wind}} \times P_t^{\text{wind}} + C_t^{\text{PV}} \times P_t^{\text{PV}} \end{array} \right\} \tag{6.22}$$

Subject to:

$$z + q_t \geq \widehat{\lambda}_t . y_t \cdot \forall t \mid \widehat{\lambda}_t > 0 \tag{6.23}$$

$$q_t \geq 0 \cdot \forall t \mid \widehat{\lambda}_t > 0 \tag{6.24}$$

$$y_t \geq 0 \tag{6.25}$$

$$z \geq 0 \tag{6.26}$$

$$P_t^p \leq y_t \tag{6.27}$$

$$\text{Constraints}(6.2) - (6.21) \tag{6.28}$$

6.4.2 Robust-Stochastic Formulation

In order to model the uncertainty of load demand and power output of WT and PV systems, the provided robust optimization model (6.22)–(6.28) is extended to consider a set of scenarios for the abovementioned uncertainties using the stochastic programming. The robust-stochastic formulation of the problem has the following form:

$$\begin{aligned} \text{Min}\,\text{OF} = {} & z_{t,s} \cdot \Gamma_t + \sum_{t=1}^{24} q_{t,s} + \sum_{l=1}^{B} \sum_{t=1}^{T} \lambda_{l,t} P_{l,t}^{\text{BC}} + \sum_{t=1}^{T} \sum_{s=1}^{Ns} P_s \\ & \times \left\{ \begin{array}{l} \lambda_{t,s} P_{t,s}^p + \sum_{j=1}^{N_j} \sum_{h=1}^{N_h} S_{j,h}^{\text{DG}} P_{j,h,t,s} \\ +C_t^B \times \left(\chi \times P_{t,s}^{\text{charge}} + \dfrac{P_{t,s}^{\text{disc}}}{\eta} \right) \\ +C_t^{\text{wind}} \times P_{t,s}^{\text{wind}} + C_t^{\text{PV}} \times P_{t,s}^{\text{PV}} \end{array} \right\} \end{aligned} \tag{6.29}$$

$$\sum_{l=1}^{B} P_{l,t}^{\text{BC}} + P_{t,s}^p + \sum_{j=1}^{N_j} \sum_{h=1}^{N_h} P_{j,h,t,s} + P_{t,s}^{\text{wind}} + P_{t,s}^{\text{PV}} + P_{t,s}^{\text{disc}} = \text{load}_{t,s}^0 + P_{t,s}^{\text{charge}}; \forall s,t \tag{6.30}$$

$$P_{l,t}^{\text{min}} s_l < P_{l,t}^{\text{BC}} < P_{l,t}^{\text{max}} s_l; \forall l,t \tag{6.31}$$

$$P_t^{\text{BC}} = \sum_{l=1}^{N_l} P_{l,t}^{\text{BC}}; \forall t \tag{6.32}$$

$$\text{Cost} = \sum_{t=1}^{T} \sum_{j=1}^{N_j} \sum_{h=1}^{N_h} \sum_{s=1}^{Ns} P_s \times S_{j,h}^{\text{DG}} P_{j,h,t,s} \tag{6.33}$$

$$0 \leq P_{j,h,t,s} \leq \left(P_{j,h}^{\text{MAX}} - P_{j,h-1}^{\text{MAX}} \right) \times U_{j,t}; \forall j,h,t,s \tag{6.34}$$

$$0 \leq P_{j,l,t,s} \leq P_{j,1}^{\text{MAX}} \times U_{j,t}; \forall j,t,s \tag{6.35}$$

$$\sum_{h=1}^{N_h} P_{j,h,t} - \sum_{h=1}^{N_h} P_{j,h,t-1} \leq R_j^{\mathrm{up}} \times U_{j,t}; \forall j, t \tag{6.36}$$

$$\sum_{h=1}^{N_h} P_{j,h,t-1} - \sum_{h=1}^{N_h} P_{j,h,t} \leq R_j^{\mathrm{down}} \times U_{j,t-1}; \forall j, t \tag{6.37}$$

$$\begin{aligned} U_{j,t} - U_{j,t-1} &\leq U_{j,t+\mathrm{Up}_{j,i}}; \forall j, \forall t, \forall i \\ U_{j,t-1} - U_{j,t} &\leq 1 - \mathrm{U}_{j,t+\mathrm{Dn}_{j,i}}; \forall j, \forall t, \forall i \end{aligned} \tag{6.38}$$

$$\begin{aligned} \mathrm{Up}_{j,i} &= \begin{Bmatrix} i & i \leq \mathrm{MUT}_j \\ 0 & i > \mathrm{MUT}_j \end{Bmatrix} \\ \mathrm{Dn}_{j,i} &= \begin{Bmatrix} i & i \leq \mathrm{MDT}_j \\ 0 & i > \mathrm{MDT}_j \end{Bmatrix} \end{aligned} \tag{6.39}$$

$$P_{t,s}^{\mathrm{wind,max}} = \begin{cases} 0 & V_{t,s}^{w} < V_{\mathrm{ci}} \\ p_r \times \left(\frac{V_{t,s}^{w} - V_{\mathrm{ci}}}{V_r - V_{\mathrm{ci}}}\right)^3 & V_{\mathrm{ci}} < V_{t,s}^{w} < V_{\mathrm{cr}} \\ p_r & V_r < V_{t,s}^{w} < V_{c0} \\ 0 & V_{t,s}^{w} > V_{c0} \end{cases} \tag{6.40}$$

$$P_{t,s}^{\mathrm{wind}} \leq P_{t,s}^{\mathrm{wind,max}}; \forall \mathrm{t, s} \tag{6.41}$$

$$P_{t,s}^{M,\max} = \frac{G_{t,s}^{a}}{G_{a_0}} \times \left\{ P_{\mathrm{Max},0}^{M} + \mu_{P_{\max}} \times \left(T_{t,s}^{a} + G_{t,s}^{a} \times \frac{\mathrm{NOCT} - 20}{800} - T_{M,0} \right) \right\} \tag{6.42}$$

$$P_{t,s}^{\mathrm{PV}} \leq P_{t,s}^{M,\max}; \forall t, s \tag{6.43}$$

$$X_{t_0}^{b} = X_0^{b} \tag{6.44}$$

$$P_{t,s}^{\mathrm{charge}} \leq P_{\mathrm{charge}}^{\max} \times U_{t,s}^{\mathrm{charge}}; \forall t, s \tag{6.45}$$

$$P_{t,s}^{\mathrm{disc}} \leq P_{\mathrm{disc}}^{\max} \times U_{t,s}^{\mathrm{disc}}; \forall t, s \tag{6.46}$$

$$X_b^{\min} \leq X_{t,s}^{b} \leq X_b^{\max}; \forall t, s \tag{6.47}$$

$$U_{t,s}^{\mathrm{charge}} + U_{t,s}^{\mathrm{disc}} \leq 1; \forall t, s \tag{6.48}$$

$$X_{t,s}^{b} = X_{t-1,s}^{b} + \chi \times P_{t,s}^{\mathrm{charge}} - \frac{P_{t,s}^{\mathrm{disc}}}{\eta}; \forall t, s \tag{6.49}$$

$$z_{t,s} + q_{t,s} \geq \widehat{\lambda}_t . y_{t,s} \cdot \forall t \mid \widehat{\lambda}_t > 0; \quad \forall \mathrm{s} \tag{6.50}$$

$$q_{t,s} \geq 0 \cdot \forall t \mid \widehat{\lambda}_t > 0; \cdot \forall \mathrm{s} \tag{6.51}$$

$$y_s \geq 0 \quad \forall \mathrm{s} \tag{6.52}$$

$$z_{t,s} \geq 0 \tag{6.53}$$

$$P^{p}_{t,s} \leq y_{t,s} \tag{6.54}$$

In the problem (6.29)–(6.54), procured power from the bilateral contracts is modeled by (6.31)–(6.32), the procured power from the self-generating units by (6.33)–(6.39), generated power from the wind turbine by (6.40)–(6.41), power production of PV by (6.42)–(6.43), and the energy storage system by (6.44)–(6.49). Also, it should be noted that the constraints (6.50)–(6.54) are imposed by the robust optimization method.

6.5 Algorithm of the Robust-Stochastic Approach

In order to solve the power procurement problem of a large consumer using the robust-stochastic model, the following steps should be taken [23]:

1. Developing the robust-stochastic model considering the operational characteristics of the large consumer and uncertain parameters (in this case, market price, load demand, wind speed, and solar irradiation)
2. Determining the conservatism level of the solution through the value of the robustness parameter Γ_t
3. Solving the robust-stochastic problem
4. Deriving the value of the uncertainty parameters out of the optimal solution obtained in step (3)

6.6 Case Study

The robust-stochastic approach is developed based on the considered sample large consumer in Chap. 2. In the following sections, obtained results are presented and discussed for the alternative sources including pool market, bilateral contracts, self-generating units, photovoltaic system, wind turbine [24], and energy storage system.

In should be noted that all the required information for the simulation can be found in Chap. 2.

6.7 Numerical Results

The developed robust-stochastic model is solved by taking 0.1 step size for the robustness parameter Γ, considering 11 iterations, and starting from zero to one.

By assigning zero to the robustness parameter, results of the stochastic method will be obtained which in this chapter is presented for scenario 50. For other values of the robustness parameter, between zero and one, the robust-stochastic case results are obtained.

6.7.1 Total Power Procurement Cost

Figure 6.2 presents the total cost of the power procurement process beside cost of procured power from different sources as pool market, bilateral contacts, and self-generating units against the robustness parameter gamma. It is obvious that in the presence of power price uncertainty, most of the required demand is obtained from self-generating units. Also, procured power from the bilateral contracts are same for different robustness parameters due to predetermined prices. Finally, it should be noted that 30% increase of the power price in the pool market is considered to solve the robust-stochastic model.

According to Fig. 6.2, it is by assigning zero for the robustness parameter that only stochastic programming results will be obtained without considering power price uncertainty. In this case obtained results for cost of purchased power from the pool market, bilateral contracts, self-generating units, and total procurement cost are

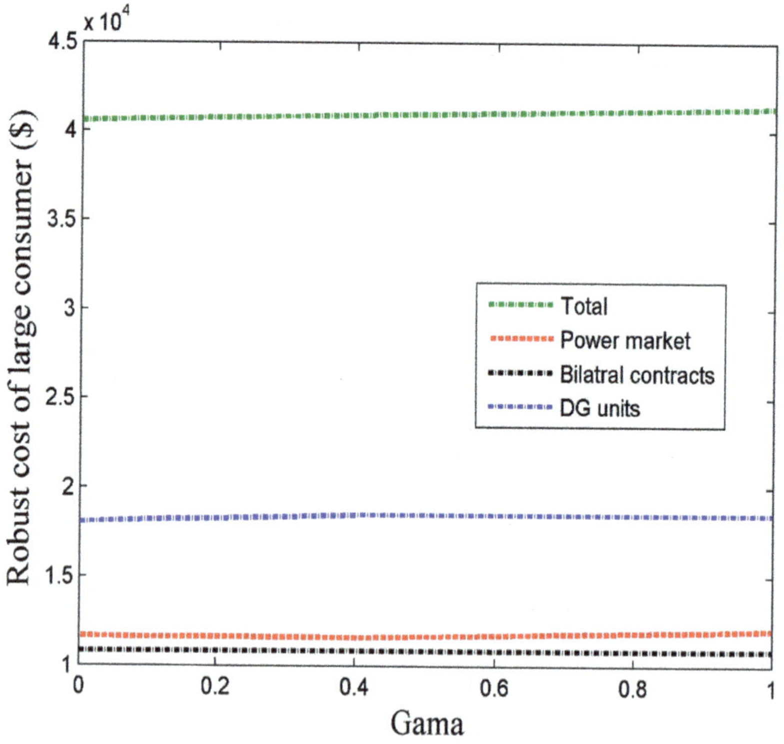

Fig. 6.2 The power procurement cost against the robustness parameter

reported as $11,680, $10,840, $18,070, and $40,590, respectively. By increasing the robustness parameter, the power procurement cost is increased due to power price uncertainty. In the worst-case simulation, which is obtained assigning one for the robustness parameter, the results are reported as $12,020, $10,840, $18,480, and $41,340 for the pool market, bilateral contracts, self-generating units, and total procurement cost, respectively. It is obvious that purchased power from the bilateral contracts has not changed due to predetermined prices. Table 6.1 provides a summary of obtained results.

Comparing results show 1.85% increase is resulted considering 30% increase in the power price uncertainty in the pool market. According to obtained results, the large consumer is highly robust against the power price uncertainty in the market because of signing bilateral contract with predetermined prices and high capacity of self-generating units.

Purchased power from the pool market is illustrated in Fig. 6.3 for stochastic and robust-stochastic cases. The results are presented for scenario 50 in the stochastic case. Also, in the robust-stochastic case, the results are presented for the worst case assuming one for the robustness parameter gamma.

The maximum purchased power from the pool market is recorded at hour 15 equal to 20.6 MW in the robust-stochastic case. In addition, the minimum purchased power is 0 at hours 1–3 and 24. In the stochastic case, the maximum and minimum purchased powers are 16.3 and 0 MW at hours 13 and 1–3, respectively.

According to Fig. 6.3, total procured power in the stochastic case is 140.9 MWh for the time horizon of the study, while this number is equal to 136.5 MWh in the robust-stochastic model. As was expected, these numbers show that procured power from the pool market is reduced due to uncertainty in the power price in the pool market.

Figure 6.4 illustrates the procured power from self-generating units during 24 h. In the stochastic solution, the maximum procured power is reported 24.4 MW at hour 21, which it is almost same for the robust-stochastic case. The minimum procured power for both cases is equal to 5.6 MW, which is recorded at hour 12.

The total procured power from the self-generating units is equal to 317.5 and 322 MWh for stochastic and robust-stochastic cases, respectively. As was expected, these numbers show that procured power from the self-generating units is increased in robust-stochastic case in comparison with stochastic case due to power price uncertainty. In other words, the operator of the large consumer decides to purchase some required demand from the self-generating units which have no uncertainty.

Table 6.1 Summary of the obtained results

Procurement cost ($)	Stochastic programming	Robust-stochastic
Pool market cost	11,680	12,020
Bilateral contracts cost	10,840	10,840
Self-generating units cost	18,070	18,480
Total procurement cost	40,590	41,340

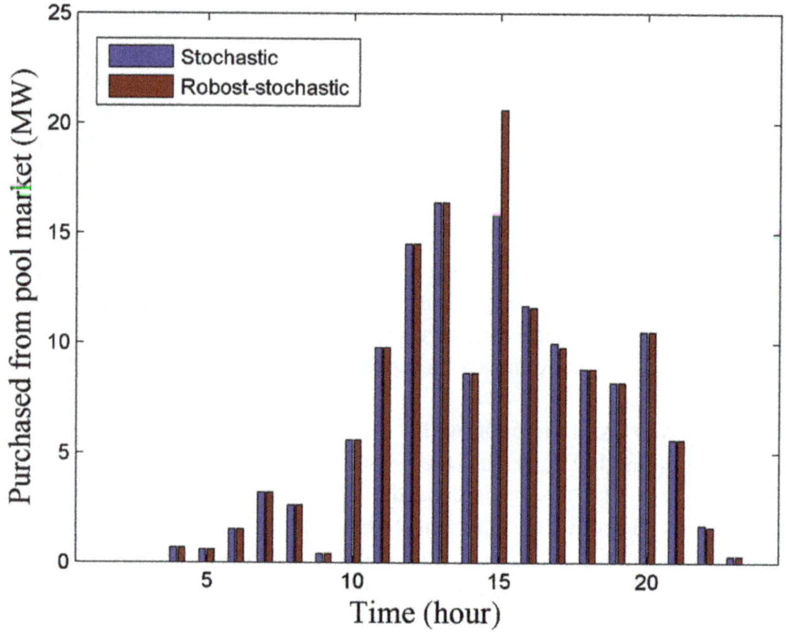

Fig. 6.3 Purchased power from the pool market

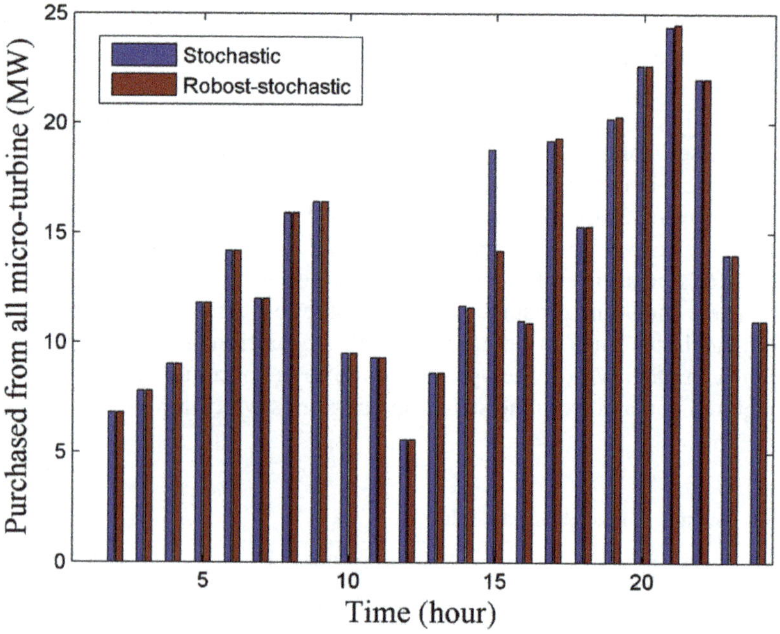

Fig. 6.4 Purchased power from the self-generating units

The purchased power from the bilateral contracts is depicted in Fig. 6.5. The procured power from the bilateral contracts is same in both cases. It should be noted that obtained result in this chapter is in accordance with the previous chapters.

According to Fig. 6.5, the maximum purchased power from the bilateral contracts is equal to 9 MW for both cases which is recorded at hours 18–24. Furthermore, purchased power at hours 1–17 is equal to 4 MW for stochastic and robust-stochastic cases. Finally, the total procured power is equal to 136 MWh during 24 h for both case studies.

The battery storage system is used to support some part of required load demand of the large consumer. Charging/discharging state of the battery is shown in Fig. 6.6 for 24 h. It should be noted that due to small capacity of the battery, its contribution is not significant in supplying the large consumer. Note that the amount of charged/ discharged power is limited for each period.

According to Fig. 6.6, there is a slight difference in the charging and discharging states of the battery storage system. For example, the battery is charged at hour 10 and discharged at hour 15 which in these periods the battery did not charge or discharge. In addition, it is obvious that the total amount of discharged power in the stochastic cases is slightly higher than the robust-stochastic case.

Figure 6.7 presents the stored energy in the battery storage system. The maximum stored energy in the battery is 2 MWh for both cases. On the other hand, the minimum stored energy is equal to 0.2 MWh. In addition the total stored energy in the storage is equal to 29.6 and 29.3 MWh for robust-stochastic and stochastic cases.

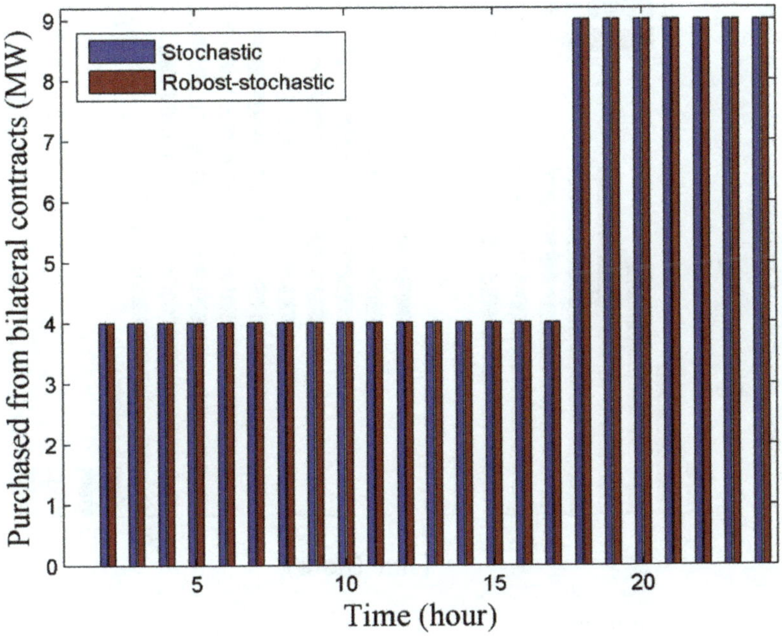

Fig. 6.5 Purchased power from the bilateral contracts

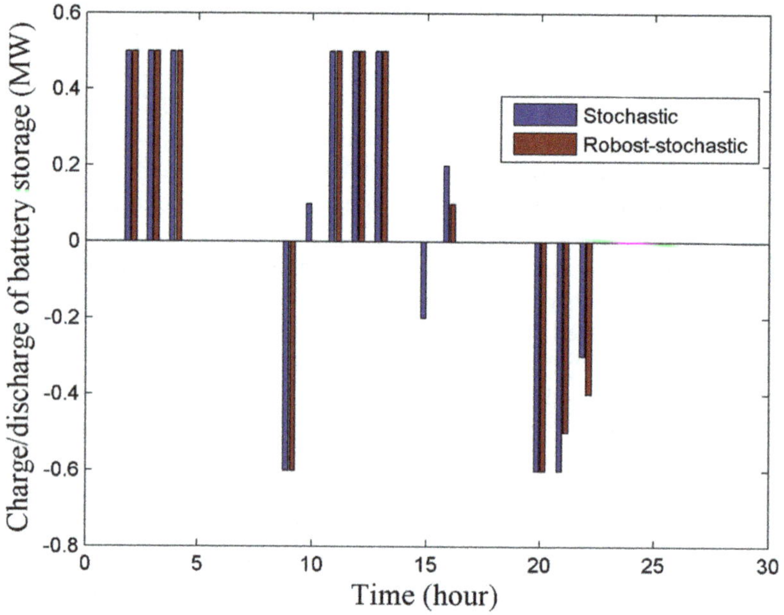

Fig. 6.6 Charging/discharging of the battery

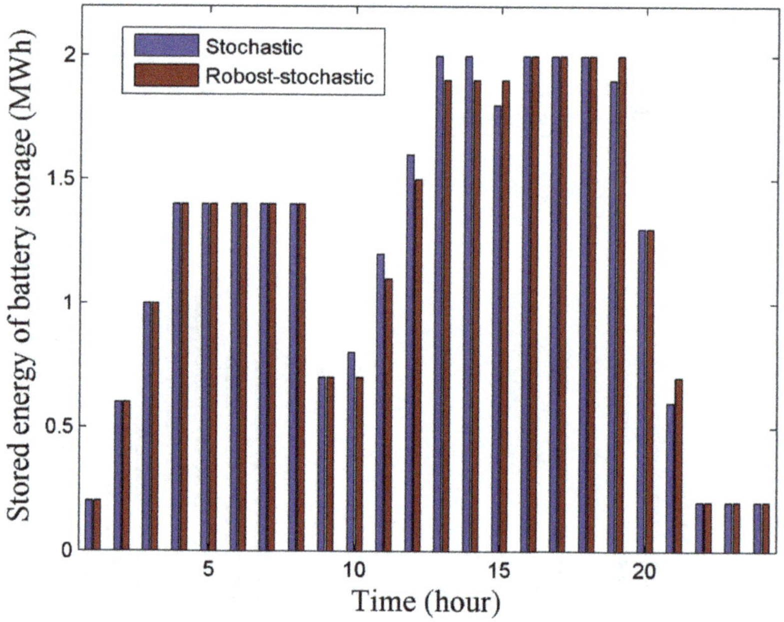

Fig. 6.7 Stored energy of the battery

6.8 Conclusion

In this chapter, a novel robust-stochastic method is developed to solve the power procurement problem faced by a large consumer. Coping with disadvantages of stochastic programming and robust optimization methods was the main motivation of developing the approach. High computational burden can be referred to as the main disadvantage of the stochastic programming. On the other hand, obtained results may not guarantee the global optimal solution. While in the robust optimization method, only on uncertain parameter can be modeled. Therefore, a novel method is developed by combining both methods.

The power procurement problem faced by a large consumer is solved by considering the battery storage system, bilateral contracts with predetermined prices, and micro-turbine as self-generating unit.

In the developed approach, load demand, solar irradiation, and wind speed are modeled using the stochastic programming method. The normal distribution is used to generate the scenarios of load demand and solar irradiation, while wind speed scenarios are generated using the Weibull distribution. It should be noted that 100 scenarios are generated to investigate the impact of the abovementioned uncertainties.

The power price on the pool market is modeled using robust optimization approach. To do so, 30% variation in the power price is taken into account to propose the robust strategy against the power price uncertainty.

Obtained results are provided in two cases: stochastic and robust-stochastic methods. As their name indicates, the former presents obtained results in the stochastic programming method without considering power price uncertainty while in the latter results of hybrid robust-stochastic approach provided in the presence of power price uncertainty.

The obtained results for cost of purchased power from the pool market, bilateral contracts, self-generating units, and total procurement cost, assigning zero for the robustness parameter, are reported as \$11,680, \$10,840, \$18,070, and \$40,590, respectively. By increasing the robustness parameter, the power procurement cost is increased due to power price uncertainty.

Assigning one for the robustness parameter, which provides the worst-case simulation results, the results are reported as \$12,020, \$10,840, \$18,480, and \$41,340 for the pool market, bilateral contracts, self-generating units, and total procurement cost, respectively. It is obvious that purchased power from the bilateral contracts has not changed due to predetermined prices. These numbers show that considering 30% increase in the power price uncertainty in the pool market, the total power procurement cost of the large consumer is increased to about 1.85% which indicates that the large consumer is robust against the power price uncertainty due to high capacity of the self-generating units and signing bilateral contracts with predetermined prices.

References

1. J.R. Birge, F. Louveaux, *Introduction to Stochastic Programming* (Springer, New York, 2011)
2. M. Gheydi, A. Nouri, N. Ghadimi, Planning in microgrids with conservation of voltage reduction. IEEE Syst. J., **12**(3), 2782–2790 (2017)
3. S. Nojavan, A. Najafi-Ghalelou, M. Majidi, K. Zare, Optimal bidding and offering strategies of merchant compressed air energy storage in deregulated electricity market using robust optimization approach. Energy **142**, 250–257 (2018)
4. P. Beraldi, A. Violi, G. Carrozzino, M.E. Bruni, A stochastic programming approach for the optimal management of aggregated distributed energy resources. Comput. Oper. Res. **96**, 200–212 (2018)
5. M. Peker, A.S. Kocaman, B.Y. Kara, A two-stage stochastic programming approach for reliability constrained power system expansion planning. Int. J. Electr. Power Energy Syst. **103**, 458–469 (2018)
6. M.-C. Hu, S.-Y. Lu, Y.-H. Chen, Stochastic–multiobjective market equilibrium analysis of a demand response program in energy market under uncertainty. Appl. Energy **182**, 500–506 (2016)
7. S. Nojavan, H. Ghesmati, K. Zare, Robust optimal offering strategy of large consumer using IGDT considering demand response programs. Electr. Power Syst. Res. **130**, 46–58 (2016)
8. R.S. Ferreira, L.A. Barroso, M.M. Carvalho, Demand response models with correlated price data: A robust optimization approach. Appl. Energy **96**, 133–149 (2012)
9. M. Rahimiyan, L. Baringo, Strategic bidding for a virtual power plant in the day-ahead and real-time markets: a price-taker robust optimization approach. IEEE Trans. Power Syst. **31**(4), 2676–2687 (2016)
10. S. Nojavan, B. Mohammadi-Ivatloo, K. Zare, Robust optimization based price-taker retailer bidding strategy under pool market price uncertainty. Int. J. Electr. Power Energy Syst. **73**, 955–963 (2015)
11. P. Xiong, P. Jirutitijaroen, C. Singh, A distributionally robust optimization model for unit commitment considering uncertain wind power generation. IEEE Trans. Power Syst. **32**(1), 39–49 (2017)
12. A. Sadeghi-Mobarakeh, A. Shahsavari, H. Haghighat, H. Mohsenian-Rad, Optimal market participation of distributed load resources under distribution network operational limits and renewable generation uncertainties. IEEE Trans. Smart Grid, 1–1 (2018), https://doi.org/10.1109/TSG.2018.2830751
13. J. Ma, Z. Song, Y. Zhang, Y. Zhao, J.S. Thorp, Robust stochastic stability analysis method of DFIG integration on power system considering virtual inertia control. IEEE Trans. Power Syst. **32**(5), 4069–4079 (2017)
14. Q. Huang, Q.-S. Jia, X. Guan, Robust scheduling of EV charging load with uncertain wind power integration. IEEE Trans. Smart Grid **9**(2), 1043–1054 (2018)
15. Y. Sasaki, N. Yorino, Y. Zoka, I. Farid, Robust stochastic dynamic load dispatch against uncertainties. IEEE Trans. Smart Grid, **9**(6), 5535–5542 (2017)
16. N.-N. Yan, L.-L. Ma, B. Liu, Research on robust stochastic optimization of multi-retailer competition under demand uncertainty, in *2008 International Conference on Machine Learning and Cybernetics*, 2008, pp. 1699–1704
17. A.R. Doodman, M. Fesanghary, R. Hosseini, A robust stochastic approach for design optimization of air cooled heat exchangers. Appl. Energy **86**(7–8), 1240–1245 (2009)
18. Z. Tan et al., The optimization model for multi-type customers assisting wind power consumptive considering uncertainty and demand response based on robust stochastic theory. Energy Convers. Manag. **105**, 1070–1081 (2015)
19. L. Ji, D.X. Niu, G.H. Huang, An inexact two-stage stochastic robust programming for residential micro-grid management-based on random demand. Energy **67**, 186–199 (2014)
20. S.Z. Moghaddam, T. Akbari, Network-constrained optimal bidding strategy of a plug-in electric vehicle aggregator: A stochastic/robust game theoretic approach. Energy **151**, 478–489 (2018)

21. S. Nojavan, K. Zare, B. Mohammadi-Ivatloo, Robust bidding and offering strategies of electricity retailer under multi-tariff pricing. Energy Econ. **68**, 359–372 (2017)
22. G. He, Q. Chen, C. Kang, Q. Xia, Optimal offering strategy for concentrating solar power plants in joint energy, reserve and regulation markets. IEEE Trans. Sustain. Energy **7**(3), 1245–1254 (2016)
23. R. Dominguez, L. Baringo, A.J. Conejo, Optimal offering strategy for a concentrating solar power plant. Appl. Energy **98**, 316–325 (2012)
24. N. Ghadimi, An adaptive neuro-fuzzy inference system for islanding detection in wind turbine as distributed generation. Complexity **21**(1), 10–20 (2015)

Chapter 7
Energy Procurement via Hybrid IGDT-Stochastic Approach

Ehsan Zargin and Raouf Morsali Asl

Abstract By using information gap decision theory (IGDT), the positive and negative aspects of the uncertainty can be analyzed. However, only one uncertain parameter can be modeled by the IGDT. In this chapter, a novel hybrid IGDT-stochastic approach is introduced to model different uncertainties in the system and utilized benefits of the IGDT and stochastic programming methods. In the power procurement problem of the large consumer, uncertainty of load demand and power output of renewable sources are modeled by a set of scenarios, and power price is modeled using the IGDT method to evaluate the positive and negative aspects of power price in the market. It should be noted that the scenarios of load demand and solar irradiation are generated by normal distribution, and the Weibull distribution is used to generate the wind speed scenarios. The application of the hybrid IGDT-stochastic method is analyzed providing numerical studies.

7.1 Introduction

In order to overcome uncertainties of the stochastic programming and robust optimization method such as high computational burden [1] and limited number of uncertain parameters [2], the novel hybrid robust-stochastic programming approach is developed in Chap. 6 to solve the power procurement problem of the large consumer. In the developed method, the uncertainties of load demand, wind speed, and solar irradiation are modeled using the stochastic programming and robust optimization approach utilized to model the power price uncertainty in the pool market [3].

As discussed in the previous chapters, the robust optimization approach only surveys the negative aspect of the uncertain parameter, while the uncertainty may have positive aspects leading to windfall benefits [4]. For example, in this particular case, power price reduction in the pool market results in low power procurement cost causing significant profit for a large industrial consumer [5, 6].

E. Zargin (✉) · R. M. Asl
Young Researchers and Elite Club, Ardabil Branch, Islamic Azad University, Ardabil, Iran

S. Nojavan et al. (eds.), *Robust Energy Procurement of Large Electricity Consumers*,
https://doi.org/10.1007/978-3-030-03229-6_7

The difference between IGDT-stochastic and robust-stochastic comes from the different input data of the IGDT and robust optimization programs [7, 8]. Furthermore, the robust optimization approach only analyzes the worst case, while the IGDT can analyze the best case of uncertainty parameters, too. However, there is similarity between the methods in which both methods are categorized as the interval optimization methods and consider the worst case of the uncertainty parameter [9, 10].

In this chapter, a novel hybrid information decision gap theory (IGDT)-stochastic approach is developed to solve the power procurement problem of the large consumer considering both positive and negative aspects of the power price uncertainty and overcoming the disadvantages of IGDT and stochastic programming methods.

In the developed hybrid IGDT-stochastic approach, the power price uncertainty is modeled using IGDT method, and the uncertainties of load demand and power output of the renewable energy sources as wind turbine and photovoltaic systems are modeled using the stochastic programming method. In order to reduce power procurement cost of the large consumer, the time-of-use rate of demand response program is implemented in the IGDT-stochastic approach. The hybrid IGDT-stochastic structure of the understudied large consumer is depicted in Fig. 7.1. Note that among different introduced IGDT models, the fractional error model is selected to solve the problem taking characteristics of the problem into account.

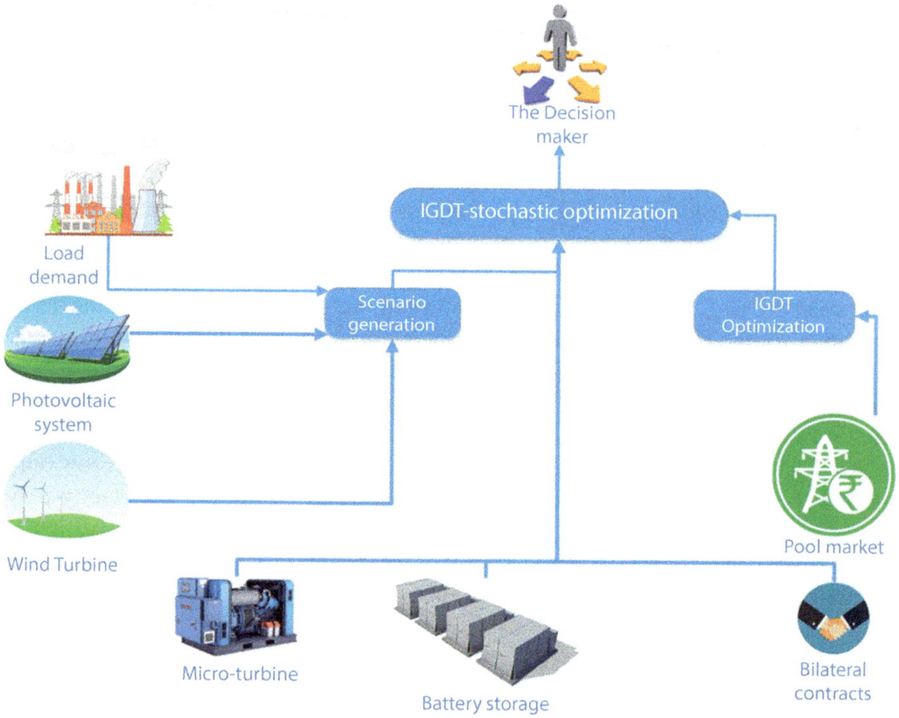

Fig. 7.1 IGDT-stochastic structure to solve power procurement problem of the large consumer

7.1.1 Brief Review of the Hybrid IGDT-Stochastic Approach

In this subsection, a brief review of the hybrid IGDT-stochastic approach is presented. This relatively new uncertainty modeling approach is used to solve various problems in the power system. In Ref. [11], the problem of determining the selling price for an electricity retailer is studied using the hybrid IGDT-stochastic approach to model the uncertainty of the problem. In this Ref., the IGDT is used to assess the opportunity and robustness decisions of optimal offering and bidding strategies in the presence of market price uncertainty, and stochastic programming is utilized to model the uncertainty of the variable climate condition containing temperature and client group demand, wind speed, and solar irradiation.

The scheduling problem of energy hub system including combined heat and power units, auxiliary boilers and energy storage devices, and wind turbine is solved in [12] through the hybrid IGDT-stochastic approach. In this Ref., the stochastic programming method is used to model the uncertainties of load demand and wind power generation, and power price uncertainty is modeled using the IGDT approach.

The optimal scheduling problem of renewable-based microgrid including micro-turbines (MTs), photovoltaic systems, wind turbines, energy storage system, and the upstream grid is solved in [13] using the hybrid IGDT-stochastic programming approach under demand response program. The stochastic programming method is used to model uncertainty modeling of load demand and output power of renewable energy sources by generating a set of scenario, and the IGDT approach is used to model the uncertainty of the electricity price.

In [5], the hybrid IGDT-stochastic approach is used to obtain the optimal bidding strategy in the day-ahead market for a large consumer. The stochastic programming method is used to model uncertainty of hourly market prices which are assumed normally distributed, and the IGDT is used to model the uncertainty of procurement cost.

7.2 Deterministic Formulation of the Problem

The power procurement problem faced by the large consumer is solved in Chap. 2 in the deterministic case considering pool market, self-generating units, bilateral contracts with predetermined price and periods, energy storage systems, wind turbine, and photovoltaic system as six alternative power sources. To develop the hybrid IGDT-stochastic approach, determinist formulation of the problem is required which has been presented in (2.1)–(2.27). To simplify the deterministic case, formulation is provided as follows:

$$\underset{P_t^p, P_{l,t}^{\mathrm{BC}}, P_{j,h,t}, P_t^{\mathrm{charge}}, P_t^{\mathrm{disc}}, P_t^{\mathrm{wind}}, P_t^{\mathrm{PV}}}{\text{Minimize}} R(q,\lambda) = \sum_{l=1}^{B}\sum_{t=1}^{T}\lambda_{l,t}P_{l,t}^{\mathrm{BC}} + \sum_{t=1}^{T}\left\{\begin{array}{l} \lambda_t P_t^p + \sum_{j=1}^{N_j}\sum_{h=1}^{N_h} S_{j,h}^{\mathrm{DG}} P_{j,h,t} \\ + C_t^B \times \left(\chi \times P_t^{\mathrm{charge}} + \frac{P_t^{\mathrm{disc}}}{\eta}\right) \\ + C_t^{\mathrm{wind}} \times P_t^{\mathrm{wind}} + C_t^{\mathrm{PV}} \times P_t^{\mathrm{PV}} \end{array}\right\} \tag{7.1}$$

$$\sum_{l=1}^{B} P_{l,t}^{\mathrm{BC}} + P_t^p + \sum_{j=1}^{N_j}\sum_{h=1}^{N_h} P_{j,h,t} + P_t^{\mathrm{wind}} + P_t^{\mathrm{PV}} + P_t^{\mathrm{disc}} = \left((1-\mathrm{DR}_t)\cdot \mathrm{load}_t^0 + \mathrm{ldr}_t\right) + P_t^{\mathrm{charge}} \tag{7.2}$$

$$P_{l,t}^{\min} s_l < P_{l,t}^{\mathrm{BC}} < P_{l,t}^{\max} s_l; \quad \forall l, t \tag{7.3}$$

$$P_t^{\mathrm{BC}} = \sum_{l=1}^{N_l} P_{l,t}^{\mathrm{BC}}; \quad \forall t \tag{7.4}$$

$$\mathrm{Cost} = \sum_{t=1}^{T}\sum_{j=1}^{N_j}\sum_{h=1}^{N_h} S_{j,h}^{\mathrm{DG}} P_{j,h,t} \tag{7.5}$$

$$0 \le P_{j,h,t} \le \left(P_{j,h}^{\mathrm{MAX}} - P_{j,h-1}^{\mathrm{MAX}}\right) \times U_{j,t}; \quad \forall j, h, t \tag{7.6}$$

$$0 \le P_{j,1,t} \le P_{j,1}^{\mathrm{MAX}} \times U_{j,t}; \quad \forall j, t \tag{7.7}$$

$$\sum_{h=1}^{N_h} P_{j,h,t} - \sum_{h=1}^{N_h} P_{j,h,t-1} \le R_j^{\mathrm{up}} \times U_{j,t}; \quad \forall j, t \tag{7.8}$$

$$\sum_{h=1}^{N_h} P_{j,h,t-1} - \sum_{h=1}^{N_h} P_{j,h,t} \le R_j^{\mathrm{down}} \times U_{j,t-1}; \quad \forall j, t \tag{7.9}$$

$$\begin{array}{l} U_{j,t} - U_{j,t-1} \le U_{j,t+\mathrm{Up}_{j,i}}; \quad \forall j, \forall t, \forall i \\ U_{j,t-1} - U_{j,t} \le 1 - U_{j,t+\mathrm{Dn}_{j,i}}; \quad \forall j, \forall t, \forall i \end{array} \tag{7.10}$$

$$\begin{array}{l} \mathrm{Up}_{j,i} = \left\{\begin{array}{ll} i & i \le \mathrm{MUT}_j \\ 0 & i \succ \mathrm{MUT}_j \end{array}\right\} \\ \mathrm{Dn}_{j,i} = \left\{\begin{array}{ll} i & i \le \mathrm{MDT}_j \\ 0 & i \succ \mathrm{MDT}_j \end{array}\right\} \end{array} \tag{7.11}$$

$$P_{t,s}^{\text{wind, max}} = \begin{cases} 0 & V_{t,s}^{w} < V_{\text{ci}} \\ p_r \times \left(\frac{V_{t,s}^{w} - V_{\text{ci}}}{V_r - V_{\text{ci}}}\right)^3 & V_{\text{ci}} < V_{t,s}^{w} < V_{\text{cr}} \\ p_r & V_r < V_{t,s}^{w} < V_{c0} \\ 0 & V_{t,s}^{w} > V_{c0} \end{cases} \tag{7.12}$$

$$P_t^{\text{wind}} \le P_t^{\text{wind, max}}; \quad \forall t \tag{7.13}$$

$$P_t^{M,\max} = \frac{G_t^a}{G_{a_0}} \times \left\{ P_{\text{Max},0}^{M} + \mu_{P\max} \times \left(T_{t,}^{a} + G_t^a \times \frac{\text{NOCT} - 20}{800} - T_{M,0} \right) \right\} \tag{7.14}$$

$$P_t^{\text{PV}} \le P_t^{M,\max}; \quad \forall t \tag{7.15}$$

$$X_{t_0}^b = X_0^b \tag{7.16}$$

$$P_t^{\text{charge}} \le P_{\text{charge}}^{\max} \times U_t^{\text{charge}}; \quad \forall t \tag{7.17}$$

$$P_t^{\text{disc}} \le P_{\text{disc}}^{\max} \times U_t^{\text{disc}}; \quad \forall t \tag{7.18}$$

$$X_b^{\min} \le X_t^b \le X_b^{\max}; \quad \forall t \tag{7.19}$$

$$U_t^{\text{charge}} + U_t^{\text{disc}} \le 1; \quad \forall t \tag{7.20}$$

$$X_t^b = X_{t-1}^b + \chi \times P_t^{\text{charge}} - \frac{P_t^{\text{disc}}}{\eta}; \quad \forall t \tag{7.21}$$

$$\text{load}_t = (1 - \text{DR}_t) \cdot \text{load}_t^0 + \text{ldr}_t \tag{7.22}$$

$$\text{load}_t^0 - \text{load}_t = \text{DR}_t \cdot \text{load}_t^0 - \text{ldr}_t \tag{7.23}$$

$$\sum_{t=1}^{T} \text{ldr}_t = \sum_{t=1}^{T} \text{DR}_t \cdot \text{load}_t^0 \tag{7.24}$$

$$\text{load}_t^{\text{inc}} \le \text{inc}_t \cdot \text{load}_t^0; \quad \forall t \tag{7.25}$$

$$\text{DR}_t \le E_{\max}; \quad \forall t \tag{7.26}$$

$$\text{inc}_t \le \text{inc}_{\max}; \quad \forall t \tag{7.27}$$

Equation (7.1) provides the total power procurement cost as the objective function. Constraint (7.2) presents the power balance between the required demand and procured power from alternative sources. The operational requirements of the problem are presented using constraints (7.3)–(7.21). Finally, constraints (7.22)–(7.27) provide the demand response program model [14].

In the model (7.1)–(7.21), $P_{l,t}^{\text{BC}}$ is the procured power from the BCs with predetermined price $\lambda_{l,t}$; P_t^P is the cost of energy procurement from the pool market with the market price λ_t; $P_{j,h,t}$ is the cost of produced power by the jth unit of the self-generation units; C_t^B is the operating cost of using battery storage, with charging/discharging efficiency χ/η and charging/discharging power of battery storage (MW) $P_t^{\text{charge}}/P_t^{\text{disc}}$; C_t^{wind} is the cost of produced power from the wind turbine; and C_t^{PV} is the cost of produced power P_t^{PV} from the PV system.

Also, $P_{l,t}^{\min}/P_{l,t}^{\max}$ is the minimum/maximum capacity relation, P_t^{BC} is the total procured power from the bilateral contracts, $P_{j,h}^{\text{MAX}}$ is the power output of block h of jth unit of self-generating units, $R_j^{\text{up}}, R_j^{\text{down}}$ is ramp-up/ramp-down rate limit of j unit, MUT_j,MDT_j is the minimum up-/downtime of self-generation units j, V_t^w is the wind speed, V_r is the rated speed, V_{ci} is the cut-in speed, V_{c0} is the cutout wind speed, $P_t^{M,\max}$ is the maximum power at any temperature T_t^a, G_t^a is the irradiation, $P_{\text{Max},0}^M$ is the maximum power, and NOCT is the normal operating cell.

In the demand response program (DRP) model (7.22)–(7.27), load_t is load after DRP performance in the same period, DR_t is the participation factor, inc_t is the size of load increase at time t, $\text{inc}_{\max}$ is the maximum size of load increase in each time t, and $\text{DR}_{\max}$ is the maximum size of load participation in DRP.

It should be noted that $\text{DR}_{\max}$ and $\text{inc}_{\max}$ are considered 20%. For more information about the formulation (7.1)–(7.27), please refer to Chap. 2.

7.3 Hybrid IGDT-Stochastic Formulation

In this section, the IGDT-stochastic approach is developed. As said before, the power price uncertainty is modeled using the IGDT optimization, and other uncertainties related to the problem are modeled using stochastic programming. In the following subsections, the IGDT-stochastic formulation of the problem is presented.

7.3.1 IGDT Formulation

In this section, the IGDT method has been applied to solve the power procurement problem faced by a large consumer considering the power price uncertainty in the pool market model. More information about the IGDT is presented in Chap. 5.

7.3.1.1 Decision Variables

The decision variables in the problem are as follows: procured power from bilateral contracts, pool market, photovoltaic systems, wind turbine, self-generating facilities, charge and discharge power, and stored energy of energy storage system.

$$q_t = \left[P_t^p, P_{l,t}^{\mathrm{BC}}, P_{j,h,t}, P_t^{\mathrm{charge}}, P_t^{\mathrm{disc}}, P_t^{\mathrm{wind}}, P_t^{\mathrm{PV}}\right]; \quad t = 1, \ldots, 24 \tag{7.28}$$

7.3.1.2 The Model of Procurement Cost

Equation (7.1) presents the cost function of electricity procurement problem faced by a large consumer which is detailed in Chap. 2.

7.3.1.3 Uncertainty Modeling

The power price in the pool market is considered as an uncertain parameter. Equation (7.29) expresses the fractional error model of info-gap method to model the uncertainty of market price.

$$U\left(\alpha, \widetilde{\lambda}_t\right) = \left\{\lambda_t : \frac{\left|\lambda_t - \widetilde{\lambda}_t\right|}{\widetilde{\lambda}_t} \leq \alpha\right\}, \alpha \geq 0 \tag{7.29}$$

where $\widetilde{\lambda}_t$ is the estimation of the pool price at time t ($/MWh) and α is the uncertainty parameter modeling the gap size between the unknown and the known.

Equation (7.29) indicates that the fractional deviation of the power price from estimated values $\widetilde{\lambda}_t$ cannot be greater than α.

7.3.1.4 Robustness Function

The maximum value of resistance against the power price uncertainty in the robustness function is shown by $\widehat{\alpha}(C_r)$. Note that in the robustness function, the high values of $\widehat{\alpha}(C_r)$ are desirable [2]. The $\widehat{\alpha}(C_r)$ can be calculated using Eq. (7.30).

$$\widehat{\alpha}(r_c) = \max_q \alpha(r_c, q) \tag{7.30}$$

To get robustness function, the maximum power procurement cost should be determined. Equation (7.29) expresses the high power prices in the market.

$$\lambda_t = \widetilde{\lambda}_t + \alpha\widetilde{\lambda}_t \tag{7.31}$$

By substituting (7.31) into (7.1) and solving for α, the robustness function can be calculated as presented in (7.32).

$$\left\{ \sum_{l=1}^{B} \sum_{t=1}^{T} \lambda_{l,t} P_{l,t}^{\text{BC}} + \sum_{t=1}^{T} \left\{ \begin{array}{l} (\widetilde{\lambda}_t + \alpha \widetilde{\lambda}_t) P_t^p + \sum_{j=1}^{N_j} \sum_{h=1}^{N_h} S_{j,h}^{\text{DG}} P_{j,h,t} \\ + C_t^B \times \left(\chi \times P_t^{\text{charge}} + \frac{P_t^{\text{disc}}}{\eta} \right) \\ + C_t^{\text{wind}} \times P_t^{\text{wind}} + C_t^{\text{PV}} \times P_t^{\text{PV}} \end{array} \right\} \right\} \leq r_c \tag{7.32}$$

$$\begin{array}{l} \text{s.t.} \\ \text{constraints}\,(7.2) - (7.27) \end{array} \tag{7.33}$$

7.3.1.5 Opportunity Function

In the IGDT method, the opportunity function as presented in (7.34) to derive risk-taker strategies. Note that the small values of $\widehat{\beta}(r_w)$ are desired [15].

$$\widehat{\beta}(r_w) = \min_q \alpha(r_w, q) \tag{7.34}$$

where r_w is cost level for the opportunity function ($).

Equation (7.35) can be used to model reduction of the power price in the pool market.

$$\lambda_t = \widetilde{\lambda}_t - \alpha \widetilde{\lambda}_t \tag{7.35}$$

Finally, the opportunity function is calculated by substituting (7.35) into (7.1) and solving for α. The resulted function is presented in (7.36).

$$\left\{ \sum_{l=1}^{B} \sum_{t=1}^{T} \lambda_{l,t} P_{l,t}^{\text{BC}} + \sum_{t=1}^{T} \left\{ \begin{array}{l} (\widetilde{\lambda}_t - \alpha \widetilde{\lambda}_t) P_t^p + \sum_{j=1}^{N_j} \sum_{h=1}^{N_h} S_{j,h}^{\text{DG}} P_{j,h,t} \\ + C_t^B \times \left(\chi \times P_t^{\text{charge}} + \frac{P_t^{\text{disc}}}{\eta} \right) \\ + C_t^{\text{wind}} \times P_t^{\text{wind}} + C_t^{\text{PV}} \times P_t^{\text{PV}} \end{array} \right\} \right\} \leq r_w \tag{7.36}$$

$$\begin{array}{l} \text{s.t.} \\ \text{constraints}\,(7.2) - (7.27) \end{array} \tag{7.37}$$

The small values of $\widehat{\beta}(r_w)$ present that the corresponding decision has low resistance against experiencing low power procurement cost levels [16].

7.3.2 IGDT-Stochastic Formulation

The uncertainty of load demand and power output of wind turbine and photovoltaic systems is modeled using the stochastic programming method. In this way, the stochastic programming is implemented on the robustness and opportunity functions separately in the following subsections.

7.3.2.1 The Robustness Function of Hybrid IGDT-Stochastic Approach

IGDT-based robustness functions (7.32) and (7.33) are updated by considering different scenarios to model load, wind speed, and solar irradiation which are resulted in (7.38)–(7.65). It should be denoted that Eqs. (7.40)–(7.65) are obtained based on Eqs. (7.2)–(7.27) by adding subscribe s to show consideration of scenarios.

$$\widehat{\alpha}(r_c) = \mathrm{Max}\alpha \tag{7.38}$$

$$\left\{ \sum_{l=1}^{B}\sum_{t=1}^{T}\lambda_{l,t}P_{l,t}^{\mathrm{BC}} + \sum_{t=1}^{T}\sum_{s=1}^{N_s}P_s \times \left\{ \begin{array}{l} (\widetilde{\lambda}_t + \alpha\widetilde{\lambda}_t)P_t^p + \sum_{j=1}^{N_j}\sum_{h=1}^{N_h} S_{j,h}^{\mathrm{DG}} P_{j,h,t,s} \\ + C_t^B \times \left(\chi \times P_{t,s}^{\mathrm{charge}} + \dfrac{P_{t,s}^{\mathrm{disc}}}{\eta}\right) \\ + C_t^{\mathrm{wind}} \times P_{t,s}^{\mathrm{wind}} + C_t^{\mathrm{PV}} \times P_{t,s}^{\mathrm{PV}} \end{array} \right\} \right\} \le r_c \tag{7.39}$$

$$\begin{aligned} &\sum_{l=1}^{B} P_{l,t}^{\mathrm{BC}} + P_{t,s}^{p} + \sum_{j=1}^{N_j}\sum_{h=1}^{N_h} P_{j,h,t,s} + P_{t,s}^{\mathrm{wind}} + P_{t,s}^{\mathrm{PV}} + P_{t,s}^{\mathrm{disc}} \\ &\quad = \left((1 - \mathrm{DR}_{t,s}) \cdot \mathrm{load}_{t,s}^{0} + \mathrm{ldr}_{t,s}\right) + P_{t,s}^{\mathrm{charge}}; \quad \forall s,t \end{aligned} \tag{7.40}$$

$$P_{l,t}^{\min} s_l < P_{l,t}^{\mathrm{BC}} < P_{l,t}^{\max} s_l; \quad \forall l,t \tag{7.41}$$

$$P_t^{\mathrm{BC}} = \sum_{l=1}^{N_l} P_{l,t}^{\mathrm{BC}}; \quad \forall t \tag{7.42}$$

$$\mathrm{Cost} = \sum_{t=1}^{T}\sum_{j=1}^{N_j}\sum_{h=1}^{N_h}\sum_{s=1}^{Ns} P_s \times S_{j,h}^{\mathrm{DG}} P_{j,h,t,s} \tag{7.43}$$

$$0 \le P_{j,h,t,s} \le \left(P_{j,h}^{\mathrm{MAX}} - P_{j,h-1}^{\mathrm{MAX}}\right) \times U_{j,t}; \quad \forall j,h,t,s \tag{7.44}$$

$$0 \le P_{j,1,t,s} \le P_{j,1}^{\mathrm{MAX}} \times U_{j,t}; \quad \forall j,t,s \tag{7.45}$$

$$\sum_{h=1}^{N_h} P_{j,h,t} - \sum_{h=1}^{N_h} P_{j,h,t-1} \le R_j^{\text{up}} \times U_{j,t}; \quad \forall j, t \tag{7.46}$$

$$\sum_{h=1}^{N_h} P_{j,h,t-1} - \sum_{h=1}^{N_h} P_{j,h,t} \le R_j^{\text{down}} \times U_{j,t-1}; \quad \forall j, t \tag{7.47}$$

$$\begin{aligned} U_{j,t} - U_{j,t-1} &\le U_{j,t+\text{Up}_{j,i}}; \quad \forall j, \forall t, \forall i \\ U_{j,t-1} - U_{j,t} &\le 1 - U_{j,t+\text{Dn}_{j,i}}; \quad \forall j, \forall t, \forall i \end{aligned} \tag{7.48}$$

$$\begin{aligned} \text{Up}_{\text{j,i}} &= \begin{Bmatrix} i & i \le \text{MUT}_j \\ 0 & i > \text{MUT}_j \end{Bmatrix} \\ \text{Dn}_{\text{j,i}} &= \begin{Bmatrix} i & i \le \text{MDT}_j \\ 0 & i > \text{MDT}_j \end{Bmatrix} \end{aligned} \tag{7.49}$$

$$P_{t,s}^{\text{wind,max}} = \begin{cases} 0 & V_{t,s}^w < V_{\text{ci}} \\ p_r \times \left(\frac{V_{t,s}^w - V_{\text{ci}}}{V_r - V_{\text{ci}}}\right)^3 & V_{\text{ci}} < V_{t,s}^w < V_{\text{cr}} \\ p_r & V_r < V_{t,s}^w < V_{c0} \\ 0 & V_{t,s}^w > V_{c0} \end{cases} \tag{7.50}$$

$$P_{t,s}^{\text{wind}} \le P_{t,s}^{\text{wind,max}}; \quad \forall t, s \tag{7.51}$$

$$P_{t,s}^{M,\max} = \frac{G_{t,s}^a}{G_{a_0}} \times \left\{ P_{\text{Max},0}^M + \mu_{P\max} \times \left(T_{t,s}^a + G_{t,s}^a \times \frac{\text{NOCT} - 20}{800} - T_{M,0} \right) \right\} \tag{7.52}$$

$$P_{t,s}^{\text{PV}} \le P_{t,s}^{M,\max}; \quad \forall t, s \tag{7.53}$$

$$X_{t_0}^b = X_0^b \tag{7.54}$$

$$P_{t,s}^{\text{charge}} \le P_{\text{charge}}^{\max} \times U_{t,s}^{\text{charge}}; \quad \forall t, s \tag{7.55}$$

$$P_{t,s}^{\text{disc}} \le P_{\text{disc}}^{\max} \times U_{t,s}^{\text{disc}}; \quad \forall t, s \tag{7.56}$$

$$X_b^{\min} \le X_{t,s}^b \le X_b^{\max}; \quad \forall t, s \tag{7.57}$$

$$U_{t,s}^{\text{charge}} + U_{t,s}^{\text{disc}} \le 1; \quad \forall t, s \tag{7.58}$$

$$X_{t,s}^b = X_{t-1,s}^b + \chi \times P_{t,s}^{\text{charge}} - \frac{P_{t,s}^{\text{disc}}}{\eta}; \quad \forall t, s \tag{7.59}$$

$$\text{load}_{t,s} = (1 - \text{DR}_{t,s}) \cdot \text{load}_{t,s}^0 + \text{ldr}_{t,s} \tag{7.60}$$

$$\text{load}_{t,s}^0 - \text{load}_{t,s} = \text{DR}_{t,s} \cdot \text{load}_{t,s}^0 - \text{ldr}_{t,s} \tag{7.61}$$

$$\sum_{t=1}^{T} \text{ldr}_{t,s} = \sum_{t=1}^{T} \text{DR}_{t,s} \cdot \text{load}_{t,s}^0; \quad \forall s \tag{7.62}$$

$$\text{load}_{t,s}^{\text{inc}} \leq \text{inc}_{t,s} \cdot \text{load}_{t,s}^{0}; \quad \forall t, s \tag{7.63}$$

$$\text{DR}_{t,s} \leq \text{DRmax}; \quad \forall t, s \tag{7.64}$$

$$\text{inc}_{t,s} \leq \text{inc}_{\max}; \quad \forall t, s \tag{7.65}$$

7.3.2.2 The Opportunity Function of Hybrid IGDT-Stochastic Approach

In order to implement stochastic model in the IGDT-based opportunity function, Eqs. (7.66)–(7.68) are obtained based on (7.36) and (7.37). These equations are updated by considering different scenarios to model price uncertainty. It should be denoted that constraints (7.40)–(7.65) are the same for both robust and opportunity functions of IGDT method.

$$\widehat{\beta}(r_w) = \text{Min}\alpha \tag{7.66}$$

$$\left\{ \sum_{l=1}^{B}\sum_{t=1}^{T}\lambda_{l,t}P_{l,t}^{\text{BC}} + \sum_{t=1}^{T}\sum_{s=1}^{N_s}P_s \times \left\{ \begin{array}{l} \left(\widetilde{\lambda}_t - \alpha\widetilde{\lambda}_t\right)P_t^p + \sum_{j=1}^{N_j}\sum_{h=1}^{N_h} S_{j,h}^{\text{DG}} P_{j,h,t,s} \\ +C_t^B \times \left(\chi \times P_{t,s}^{\text{charge}} + \dfrac{P_{t,s}^{\text{disc}}}{\eta}\right) \\ +C_t^{\text{wind}} \times P_{t,s}^{\text{wind}} + C_t^{\text{PV}} \times P_{t,s}^{\text{PV}} \end{array} \right\} \right\} \leq r_w \tag{7.67}$$

$$\begin{array}{l} \text{s.t.} \\ \text{Constraints}\,(7.40) - (7.65) \end{array} \tag{7.68}$$

7.4 Numerical Results

In this section, obtained results of robustness and opportunity of the hybrid IGDT-stochastic approach are presented in three strategies as risk-averse, risk-neutral, and risk-taker. Risk-averse and risk-taker strategies present the results of the robustness and opportunity functions, respectively, and stochastic case results are presented as risk-neutral strategy. The problem is solved in two cases as with and without considering demand response program to show the impact of time-of-use rate of demand response program. As said before, the stochastic programming method is used to model the load demand, wind speed, and solar irradiation generation using ten discrete scenarios which the Weibull distribution is used to generate scenarios for wind speed and normal distribution is used to generate scenarios for other uncertainties.

Table 7.1 Results of risk-neutral strategy

Different costs	Without DRP	With DRP
Procurement cost from pool market ($)	13,377.977	11,183.678
Procurement cost from BCs ($)	9322.080	9322.080
Procurement cost from SG units ($)	18,023.477	18,309.788
Total procurement cost ($)	40,723.534	38,815.545

7.4.1 *Risk-Neutral Results*

The risk-neutral strategy is obtained without considering power price uncertainty in the pool market, assigning zero for robustness or opportunity parameter in the robustness or opportunity function. In this strategy, total power procurement cost is equal to $40,723.534 and $38,815.545 for without and with implementing demand response programs. A summary of obtained results is provided in Table 7.1. It is obvious that by implementing the TOU-DRP, total power procurement cost is reduced to about 4.7%.

7.4.2 *Risk-Averse Results*

Risk-averse strategy is developed to consider the pernicious effects of the uncertainty using the robustness function in the IGDT method. Figure 7.2 presents the robustness procurement cost against the robustness function $\widehat{\alpha}(r_c)$ for without and with considering DRP.

It is obvious that by increasing the robustness cost r_c, the robustness is increased which was expected. This means that by paying higher procurement cost, higher robustness against the power price uncertainty can be attained. In the without DRP case, by paying $42,000 to procure the electricity demand, 13.4% robustness against the power price uncertainty can be obtained. Same amount of uncertainty is achieved by spending $40,000.

7.4.3 *Risk-Taker Results*

The risk-taker strategy that is derived using the opportunity function is developed in the IGDT method to model the positive sides of the uncertainty. The opportunity function results considering power price reduction in the pool market are depicted in Fig. 7.3 in the IGDT-stochastic approach. According to Fig. 7.3, for 15% power price reduction, the power procurement cost is equal to $38,320 and $36,700 for without and with DRP cases, indicating 4.22% reduction because of implementing DRP.

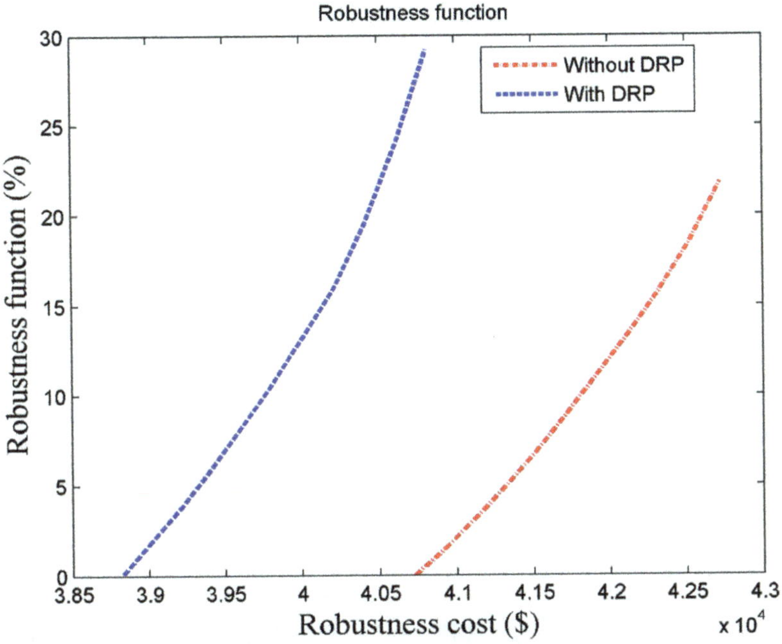

Fig. 7.2 Robustness cost for without and with considering DRP

7.4.4 *Analyzing of Alternative Sources of the Large Consumer*

The impact of power price uncertainty on the alternative power sources is investigated in the following subsection through three strategies as risk-taker, risk-neutral, and risk-averse. Note that all results are reported for scenario 5 in the hybrid IGDT-stochastic approach.

7.4.4.1 Load Profile

Figure 7.4 illustrates the new load profile in the three strategies for without and with implementing DRP in scenario 5. The TOU-DRP has shifted the load from peak periods to off-peak periods resulting lower procurement. In the without DRP case, the maximum required demand is reported as 41.5 MW at hour 20 for three considered strategies. After applying DRP, the maximum required demand is recorded as 36.5 MW at hour 13 for the risk-neutral strategy and 47.5 MW for the other two strategies.

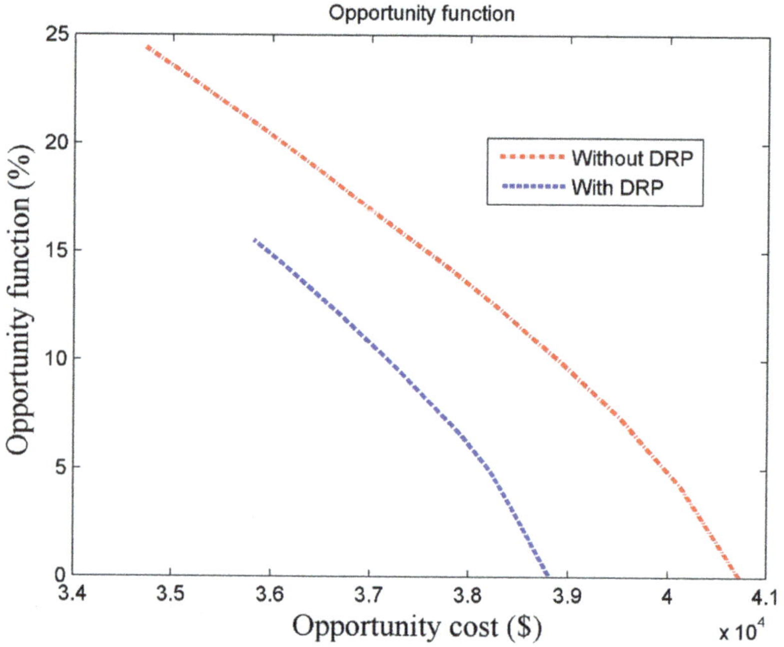

Fig. 7.3 Opportunity cost for without and with considering DRP

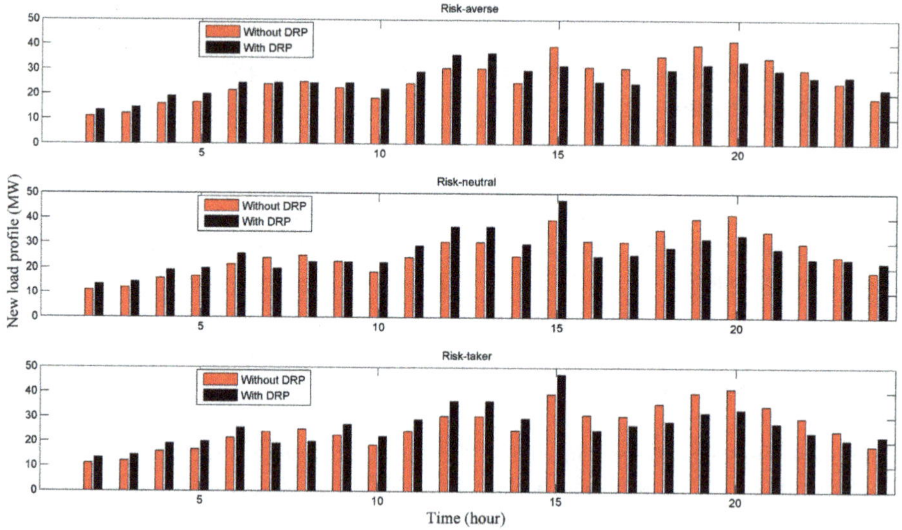

Fig. 7.4 New load profile without and with implementing DRP for risk-averse, risk-neutral, and risk-taker strategies

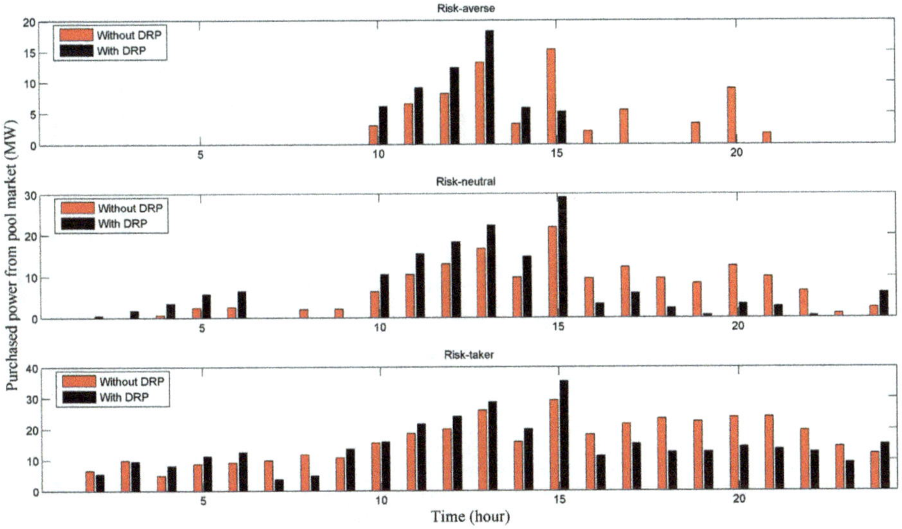

Fig. 7.5 Purchased power from the pool market in the three strategies

7.4.4.2 Purchased Power from the Pool Market

The purchased power from the pool market is shown in Fig. 7.5 for time horizon of the study. It should be reminded that power price increase is considered in the risk-averse strategy and risk-taker strategy considers the power price reduction. Therefore, purchased power in the risk-averse strategy is reduced to be robust against the power price uncertainty. On the other hand, in the risk-taker strategy, purchased power is increased due to less power prices in the market.

To supply high load demand during the high load demand periods 10–16, the amount of purchased power is almost same. Moreover, it is obvious that the purchased power from the pool market is reduced due to the implementation of DRP.

7.4.4.3 Purchased Power from the Bilateral Contracts

Procured power from the bilateral contracts is depicted in Fig. 7.6 for risk-averse, risk-neutral, and risk-taker. According to Fig. 7.6, in the risk-averse strategy, more power is procured in comparison with the other strategies. In addition, by purchasing power from the pool market in the low power price situation to supply the load demand, the procured power in the risk-taker strategy is equal to zero.

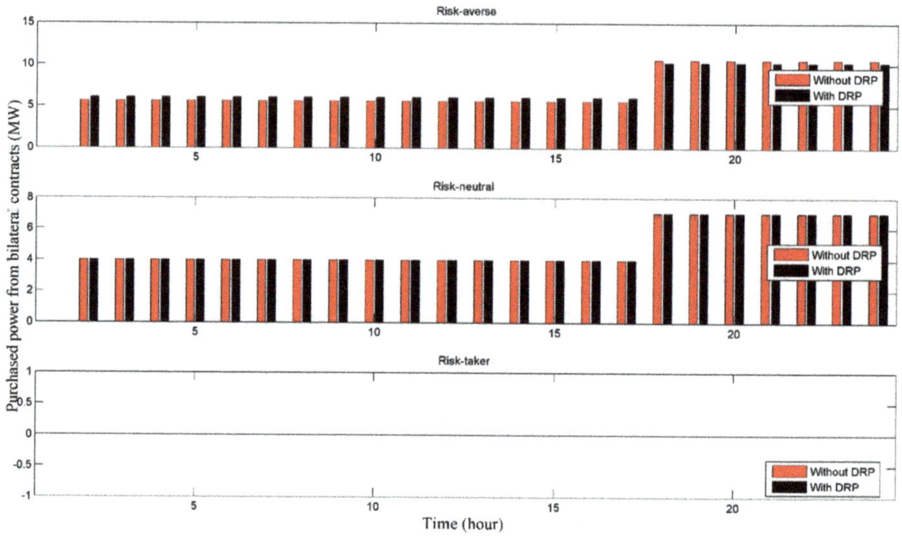

Fig. 7.6 Procured power from the bilateral contracts units in the three strategies

7.4.4.4 Purchased Power from the Self-Generating Units

Figure 7.7 shows the purchased power from self-generating units for the three strategies. It is obvious that purchased power in the risk-averse strategy is higher than others to be robust against power price increase in the pool market. In addition, the impact of the DRP on the purchased power from micro-turbines is quite moderate.

7.4.4.5 Procured Power from the Storage Systems

Finally, Figs. 7.8 and 7.9 illustrate the charge/discharge state and stored energy in the battery for the three strategies. The number and time of charging/discharging of the battery storage in the risk-averse strategy are less than the risk-neutral and risk-taker strategies.

According to Fig. 7.9, the maximum stored energy in the battery is recorded in without DRP case of risk-averse strategy equal to 28.5 MWh. This number is reduced to 22.5 MWh after implementing DRP. The total stored energy in the risk-neutral strategy is reported as 30.10 and 22.5 MWh for without and with DRP cases, respectively. Finally, in the risk-averse strategy, these numbers are reported as 30.3 and 24 MWh for without and with DRP cases, respectively. According to the abovementioned descriptions, implementing DRP has reduced stored energy of the battery in the all of the considered strategies.

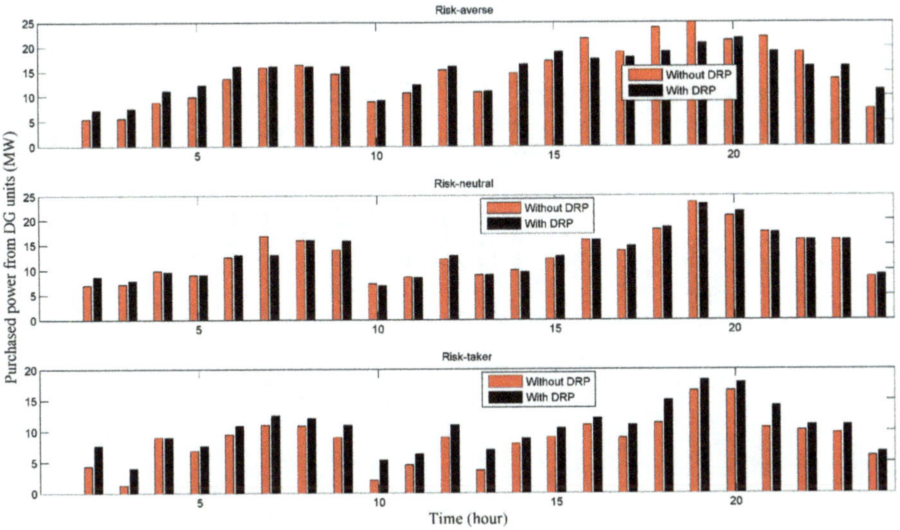

Fig. 7.7 Procured power from the self-generating unit

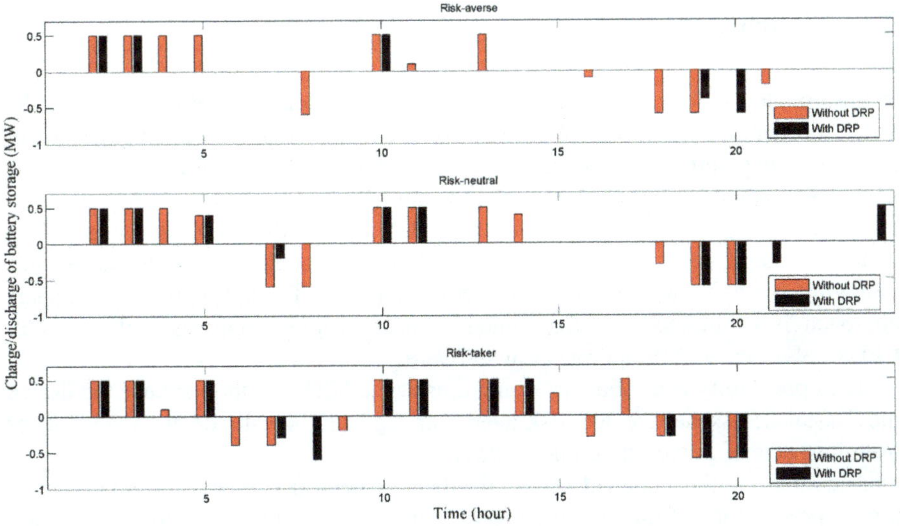

Fig. 7.8 Charging/discharging of the battery in the three strategies

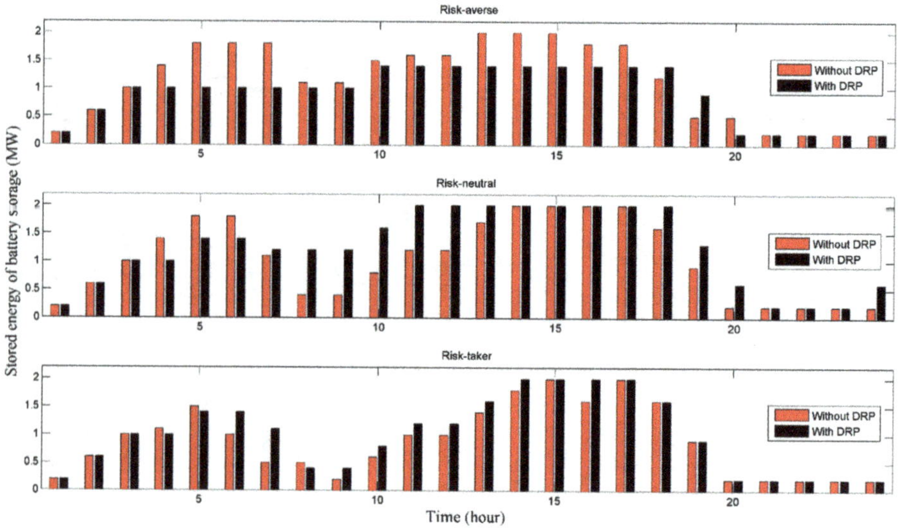

Fig. 7.9 Stored energy in the battery in the three strategies

7.5 Conclusion

In this chapter, a novel hybrid IGDT-stochastic approach is developed to solve the power procurement problem of a large consumer under the uncertainty of power price, load demand, wind speed, and solar irradiation. The main advantage of the developed method comes from modeling multiple uncertain parameters simultaneously. The power price uncertainty is modeled using the IGDT and uncertainty of load demand, solar irradiation, and wind speed modeled using stochastic programming method via generating ten discrete scenarios. The hybrid IGDT-stochastic approach is formulated as mixed-integer linear programming and solved under CPLX solver in GAMS optimization software.

The opportunity and robustness functions of the IGDT method are used to derive the risk-taker, risk-neutral, and risk-averse strategies to investigate the pernicious or useful aspects of power price uncertainty.

According to the obtained results, in the risk-neutral strategy, the total cost of power procurement from the alternative sources is obtained as \$40,723.534 and \$38,815.545 for without and with considering demand response programs, respectively, which means by implementing the TOU-DRP, the total power procurement cost has reduced to about 4.7%.

In the risk-averse strategy, by increasing the robustness cost, the robustness of the large consumer against the power price uncertainty is increased. As an example, by paying \$42,000 in the without DRP case, which is about 1.1% higher than risk-neutral cost, the large consumer is robust against the 13.4% of the power price increase. The same amount of uncertainty is achieved by spending \$40,000.

In the risk-taker strategy, which is derived to model the positive aspect of the uncertainty, for 15% reduction in the power price, the power procurement cost is equal to \$38,320 and \$36,700 for without and with DRP cases, indicating 4.22% reduction because of implementing DRP.

References

1. S. Nojavan, H. allah Aalami, Stochastic energy procurement of large electricity consumer considering photovoltaic, wind-turbine, micro-turbines, energy storage system in the presence of demand response program. Energy Convers. Manag. **103**, 1008–1018 (2015)
2. S. Nojavan, H. Ghesmati, K. Zare, Robust optimal offering strategy of large consumer using IGDT considering demand response programs. Electr. Power Syst. Res. **130**, 46–58 (2016)
3. N. Ghadimi, M.H. Firouz, Short-term management of hydro-power systems based on uncertainty model in electricity markets. J. Power Technol. **95**(4), 265–272
4. S. Nojavan, K. Zare, M.A. Ashpazi, A hybrid approach based on IGDT–MPSO method for optimal bidding strategy of price-taker generation station in day-ahead electricity market. Int. J. Electr. Power Energy Syst. **69**, 335–343 (2015)
5. K. Zare, M.P. Moghaddam, M.K. Sheikh El Eslami, Demand bidding construction for a large consumer through a hybrid IGDT-probability methodology. Energy **35**(7), 2999–3007 (2010)
6. M. Tarafdar Hagh, H. Ebrahimian, N. Ghadimi, Hybrid intelligent water drop bundled wavelet neural network to solve the islanding detection by inverter-based DG. Front. Energy **9**(1), 75–90 (2015)
7. A. Najafi-Ghalelou, S. Nojavan, K. Zare, Information gap decision theory-based risk-constrained scheduling of smart home energy consumption in the presence of solar thermal storage system. Sol. Energy **163**, 271–287 (2018)
8. A. Kazemi, S. Dehghan, N. Amjady, Multi-objective robust transmission expansion planning using information-gap decision theory and augmented ε-constraint method. IET Gener. Transm. Distrib. **8**(5), 828–840 (2014)
9. S. Nojavan, K. Zare, M.R. Feyzi, Optimal bidding strategy of generation station in power market using information gap decision theory (IGDT). Electr. Power Syst. Res. **96**, 56–63 (2013)
10. J. Zhao, C. Wan, Z. Xu, J. Wang, Risk-based day-ahead scheduling of electric vehicle aggregator using information gap decision theory. IEEE Trans. Smart Grid **8**(4), 1609–1618 (2017)
11. S. Nojavan, K. Zare, B. Mohammadi-Ivatloo, Risk-based framework for supplying electricity from renewable generation-owning retailers to price-sensitive customers using information gap decision theory. Int. J. Electr. Power Energy Syst. **93**, 156–170 (2017)
12. A. Dolatabadi, M. Jadidbonab, B. Mohammadi-ivatloo, Short-term scheduling strategy for wind-based energy hub: a hybrid stochastic/IGDT approach. IEEE Trans. Sustain. Energy, 1-1 (2018)
13. A. Mehdizadeh, N. Taghizadegan, J. Salehi, Risk-based energy management of renewable-based microgrid using information gap decision theory in the presence of peak load management. Appl. Energy **211**, 617–630 (2018)
14. S. Nojavan, K. Zare, B. Mohammadi-Ivatloo, Selling price determination by electricity retailer in the smart grid under demand side management in the presence of the electrolyser and fuel cell as hydrogen storage system. Int. J. Hydrog. Energy **42**(5), 3294–3308 (2017)
15. A. Najafi-Ghalelou, S. Nojavan, K. Zare, Heating and power hub models for robust performance of smart building using information gap decision theory. Int. J. Electr. Power Energy Syst. **98**, 23–35 (2018)
16. Y. Ben-Haim, Y. Ben-Haim, *Info-gap decision theory: decisions under severe uncertainty* (Academic, West Bengal, 2006)

Chapter 8
Robust Energy Procurement Under Time-of-Use Pricing

Sajjad Dadfar

Abstract The time-of-use (TOU) pricing is one of the most important demand response programs in an electricity market, which is used to manage the peak load demand. In this way, capital investment of extra power plants, which may be required few hours during a year, is eliminated. In the TOU demand response program, a constant set of tariffs for different hours of the day and/or seasons is defined at the beginning of a given horizon.

In this chapter, the TOU demand response program is considered to implement the power procurement problem of a large consumer in the presence of pool price uncertainty. The uncertainty of power price in the pool market is modeled using the robust optimization method in which the sensitivity of the optimal solution to the power price uncertainty is reduced. The application of the TOU is reported and analyzed through numerical studies. In addition, obtained results are compared with the deterministic case and without implementing TOU demand response programs.

8.1 Introduction

In the previous chapters, the power procurement problem faced by large consumer is solved in deterministic and uncertain modes using different methods as stochastic programming, robust optimization approach, IGDT method, hybrid robust-stochastic method, and IGDT-stochastic method.

In the following two chapters, the problem is solved under demand response programs to describe usefulness of such programs in detail. So, in this chapter, robust energy procurement of a large industrial consumer is pursued under time-of-use (TOU) pricing demand response program (DRP).

S. Dadfar (✉)
Department of Electrical Engineering, College of Electrical Engineering and Computer, Saveh Branch, Islamic Azad University, Saveh, Iran

Ministry of Energy, Iran Power Generation Transmission and Distribution Management Organization-Head Office (Tavanir), Tehran, Iran

S. Nojavan et al. (eds.), *Robust Energy Procurement of Large Electricity Consumers*, https://doi.org/10.1007/978-3-030-03229-6_8

The main motivation of implementing DRPs is the fact that commonly 10% of peak demand occurs within 1% of the hours in a year, usually during the summer months based on the US electric systems [1]. To supply this peak demand which is experienced only few hours during a year, significant amount of money should be invested on constructing costly extra power plants, transmission lines, and distribution systems, which in most of time are useless [2]. On the other hand, by managing the load during peak load demand, all of those costly investments and activities can be prevented. For example, according to [1], 625 peak power plants and the corresponding delivery infrastructure would be eliminated by reducing 5% of peak load demand. In addition, DRPs can be used to reduce the price volatility during peak demand hours.

Therefore, the DRPs are designed to motivate the consumers to reduce or even cut out their load when required. Currently, in spite of inadequate standards, DRPs are implemented in the US electric systems [3, 4]. Demand response programs are divided into two main categories: incentive-based programs (IBPs) and time-based programs (TBPs) [5, 6]. According to Fig. 8.1, TOU rate of DRP is classified as PBP. More information about the TOU is presented in the following subsections.

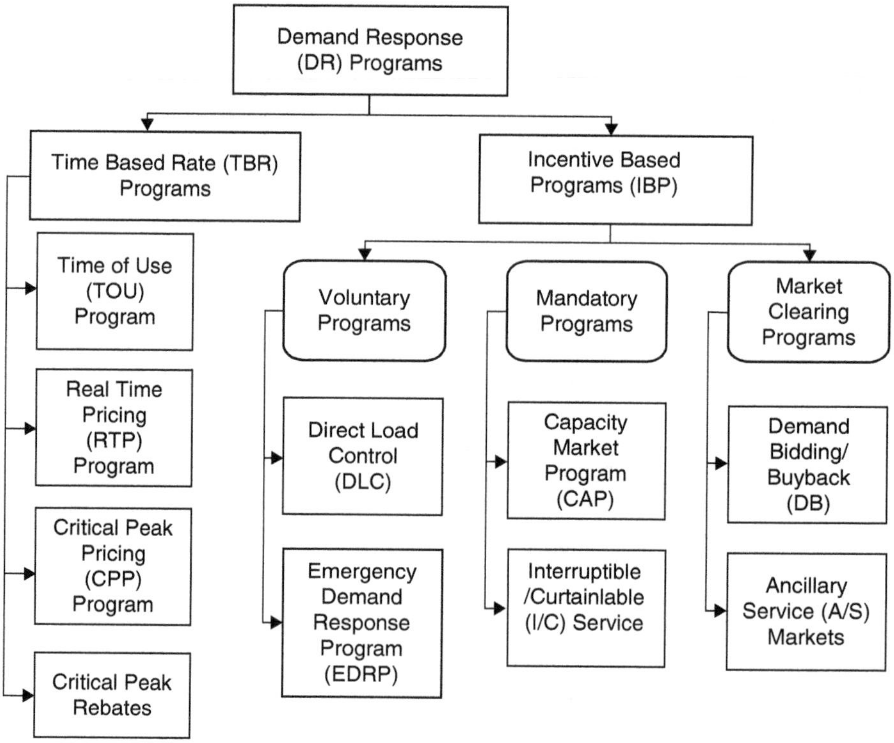

Fig. 8.1 Classification of demand response programs

8.2 Background of TOU

Using bulk storage to store electricity energy is not possible because of a lack of economical technology [7]. This means that the produced energy should be consumed instantly. Most of electricity consumers are charged a fixed rate per kilowatt-hour which is resulted in predictable load profile affected by the time of the day, day of the week, season, weather, etc. The flat-rate pricing is the most common pricing system in which the consumer pays same amount of money per unit of consumed energy without considering the time or system conditions. Note that there might be significant difference between the paid price for the flat-rate and marginal-cost pricing of electricity over a year which results in higher consumption during peak periods and less consumption during the off-peak periods, considering economical motivations. Therefore, demand-side management (DSM) programs are developed to efficiently use resources and reduce power production cost. As said before, DSMs are categorized as IBPs and TBPs. The TBPs are divided into real-time pricing, critical peak pricing, and time-of-use (TOU) pricing.

The consumer will charge more during peak periods under the TOU pricing rate of DRP [8]. In spite of real-time and critical peak pricing programs, in the TOU pricing, there is no need for two-way communications between consumers and utilities which has made TOU as a viable demand management strategy. In the TOU pricing, consumers receive a certain price for different periods of days of a week. It should be noted that in other TBPs, the consumer receives varied price at real time [9]. Theoretically, TOU pricing encourages the consumers to shift their load demand across time blocks, which prevents expansion of generation capacity [10].

Implementing the TBPs requires infrastructure augmentation using smart meters to track the time and amount of consumed energy by the consumer despite traditional electricity systems in which tracking the total consumed energy was enough over a billing period to charge the consumers. Therefore, the smart meters are essential components toward a smart grid resulting in higher efficiency of electricity systems and optimal electricity energy usage.

In the USA, the smart meters have gained support with generous federal funding [11], although these devices are expensive infrastructures which are essential for the implementation of TOU rate of DRP.

Ontario, a province in Canada, announced a policy to reduce energy consumption through developing the smart grid and TBPs, in April 2004. Later in 2006, TOU pricing program is implemented for small business and residential electricity consumers with three periods (on-peak, mid-peak, off-peak) and two seasons (November to April and May to October). Finally, in 2010, the provincial government mandated smart meter installation and the transition of TOU pricing for all households in the six provinces.

Incentivizing customers to reduce electricity consumption during the peak load and price periods and shift the demand to the lower price and load periods was the main goal of implementing the TOU rates in Ontario, which reflected the government intention to utilize the current capacity of electricity generation, by implementing electricity TBP demand management strategy in Ontario [12].

By increasing interest on smart metering devices, TBPs have received more attention including the TOU pricing.

In [12], 15 experimental studies are analyzed in which the obtained results show that the peak demand is reduced between 3 and 6% due to implementation of TOU rate of DRP. Note that all the analyzed in this Ref. are studied in a short-term period under voluntary participation.

The Ontario case is a natural experiment for evaluating the impact of electricity pricing, because it is the only region that mandate smart meters to all its residential customers. Published reports about the economic analysis of TOU pricing in Ontario and academic papers are really scant. Separate consulting studies reported in [13] indicate that the peak time demand is reduced about 3% after 1 year of mandatory TOU pricing.

8.3 Formulation of TOU

The importance of implementing TOU is detailed in the previous section. In this section the TOU rate of DRP is formulated based on the obtained model in the literature. Note that the model is used in Chap. 2 to solve the power procurement problem of a large consumer in a deterministic mode. The TOU rate of DRP transfers load from high-price periods to low-price periods [14] to reduce the energy procurement cost. According to [15], a large consumer can implement the time-of-use (TOU) rate of DRP, which has a limit on the maximum shifting capability of the load. This limit is assumed 20% in this chapter.

Equations (8.1) and (8.2) mathematically model the TOU rate of DRP.

$$\mathrm{load}_t = (1 - \mathrm{DR}_t) \cdot \mathrm{load}_t^0 + \mathrm{ldr}_t \tag{8.1}$$

$$\mathrm{load}_t^0 - \mathrm{load}_t = \mathrm{DR}_t \cdot \mathrm{load}_t^0 - \mathrm{ldr}_t \tag{8.2}$$

where

load_t is the load after DRP performance in the same period
DR_t is the participation factor of large consumer

Note that in TOU pricing, the transferred load amount highly depends on the power price in the pool market [8].

Equations (8.3)–(8.6) present the related constraints to the DRP. According to constraint (8.3), the transferable load has variable size, which is dependent on DR_t. Constraint (8.4) limits load increase at each period. The maximum amount of DR_t and inc_t in each time is limited by constraints (8.5) and (8.6). In this chapter, $\mathrm{DR}_{\max}$ and $\mathrm{inc}_{\max}$ are considered 20%.

$$\sum_{t=1}^{T} \mathrm{ldr}_t = \sum_{t=1}^{T} \mathrm{DR}_t \cdot \mathrm{load}_t^0 \tag{8.3}$$

$$\mathrm{load}_t^{\mathrm{inc}} \le \mathrm{inc}_t \cdot \mathrm{load}_t^0; \quad \forall t \tag{8.4}$$

$$\mathrm{DR}_t \le \mathrm{DR}_{\max}; \quad \forall t \tag{8.5}$$

$$\mathrm{inc}_t \le \mathrm{inc}_{\max}; \quad \forall t \tag{8.6}$$

where

inc_t is the size of load increase at time t
$\mathrm{inc}_{\max}$ is the maximum size of load increase in each time t
$\mathrm{DR}_{\max}$ is the maximum size of load participation in DRP

8.4 Robust Energy Procurement Formulation

In this chapter, the robust optimization method is used to study the impact of power price uncertainty in the pool market under TOU rate of DRP. Detailed information about the robust optimization approach is provided in Chap. 4. Therefore, in this chapter only the formulation of robust energy procurement of large consumer under the TOU pricing is presented as follows:

$$\begin{aligned} \text{Min OF} &= z \cdot \Gamma + \sum_{t=1}^{24} q_t + \sum_{l=1}^{B} \sum_{t=1}^{T} \lambda_{l,t} P_{l,t}^{\mathrm{BC}} \\ &+ \sum_{t=1}^{T} \left\{ \begin{array}{l} \lambda_t P_t^p + \sum_{j=1}^{N_j} \sum_{h=1}^{N_h} S_{j,h}^{\mathrm{DG}} P_{j,h,t} \\ + C_t^B \times \left(\chi \times P_t^{\mathrm{charge}} + \dfrac{P_t^{\mathrm{disc}}}{\eta} \right) \\ + C_t^{\mathrm{wind}} \times P_t^{\mathrm{wind}} + C_t^{\mathrm{PV}} \times P_t^{\mathrm{PV}} \end{array} \right\} \end{aligned} \tag{8.7}$$

Subject to

$$z + q_t \ge \widehat{\lambda}_t . y_t \quad \forall t \mid \widehat{\lambda}_t > 0 \tag{8.8}$$

$$q_t \ge 0 \quad \forall t \mid \widehat{\lambda}_t > 0 \tag{8.9}$$

$$y_t \ge 0 \tag{8.10}$$

$$z \ge 0 \tag{8.11}$$

$$P_t^p \le y_t \tag{8.12}$$

$$q_t \geq 0 \quad \forall t \mid \widehat{\lambda}_t > 0 \tag{8.13}$$

$$\sum_{l=1}^{B} P_{l,t}^{\mathrm{BC}} + P_t^{p} + \sum_{j=1}^{N_j}\sum_{h=1}^{N_h} P_{j,h,t} + P_t^{\mathrm{wind}} + P_t^{\mathrm{PV}} + P_t^{\mathrm{disc}} = \left((1-\mathrm{DR}_t)\cdot \mathrm{load}_t^0 + \mathrm{ldr}_t\right) + P_t^{\mathrm{charge}} \tag{8.14}$$

$$P_{l,t}^{\min} s_l < P_{l,t}^{BC} < P_{l,t}^{\max} s_l; \quad \forall l, t \tag{8.15}$$

$$P_t^{\mathrm{BC}} = \sum_{l=1}^{N_l} P_{l,t}^{\mathrm{BC}}; \quad \forall t \tag{8.16}$$

$$\mathrm{Cost} = \sum_{t=1}^{T}\sum_{j=1}^{N_j}\sum_{h=1}^{N_h} S_{j,h}^{\mathrm{DG}} P_{j,h,t} \tag{8.17}$$

$$0 \leq P_{j,h,t} \leq \left(P_{j,h}^{\mathrm{MAX}} - P_{j,h-1}^{\mathrm{MAX}}\right) \times U_{j,t}; \quad \forall j, h, t \tag{8.18}$$

$$0 \leq P_{j,1,t} \leq P_{j,1}^{\mathrm{MAX}} \times U_{j,t}; \quad \forall j, t \tag{8.19}$$

$$\sum_{h=1}^{N_h} P_{j,h,t} - \sum_{h=1}^{N_h} P_{j,h,t-1} \leq R_j^{\mathrm{up}} \times U_{j,t}; \quad \forall j, t \tag{8.20}$$

$$\sum_{h=1}^{N_h} P_{j,h,t-1} - \sum_{h=1}^{N_h} P_{j,h,t} \leq R_j^{\mathrm{down}} \times U_{j,t-1}; \quad \forall j, t \tag{8.21}$$

$$\begin{array}{l} U_{j,t} - U_{j,t-1} \leq U_{j,t+\mathrm{Up}_{j,i}}; \quad \forall j, \forall t, \forall i \\ \mathrm{U}_{j,t-1} - \mathrm{U}_{j,t} \leq 1 - \mathrm{U}_{j,t+\mathrm{Dn}_{j,i}}; \quad \forall j, \forall t, \forall i \end{array} \tag{8.22}$$

$$\begin{array}{l} \mathrm{Up}_{j,i} = \left\{\begin{array}{ll} i & i \leq \mathrm{MUT}_j \\ 0 & i \succ \mathrm{MUT}_j \end{array}\right\} \\ \mathrm{Dn}_{j,i} = \left\{\begin{array}{ll} i & i \leq \mathrm{MDT}_j \\ 0 & i \succ \mathrm{MDT}_j \end{array}\right\} \end{array} \tag{8.23}$$

$$P_{t,s}^{\mathrm{wind,\,max}} = \begin{cases} 0 & V_{t,s}^{w} < V_{\mathrm{ci}} \\ p_r \times \left(\frac{V_{t,s}^{w} - V_{\mathrm{ci}}}{V_r - V_{\mathrm{ci}}}\right)^3 & V_{\mathrm{ci}} < V_{t,s}^{w} < V_{\mathrm{cr}} \\ p_r & V_r < V_{t,s}^{w} < V_{c0} \\ 0 & V_{t,s}^{w} > V_{c0} \end{cases} \tag{8.24}$$

$$P_t^{\mathrm{wind}} \leq P_t^{\mathrm{wind,\,max}}; \quad \forall t \tag{8.25}$$

$$P_t^{M,\max} = \frac{G_t^a}{G_{a_0}} \times \left\{P_{\mathrm{Max},0}^{M} + \mu_{P\max} \times \left(T_{t,}^{a} + G_t^a \times \frac{\mathrm{NOCT} - 20}{800} - T_{M,0}\right)\right\} \tag{8.26}$$

$$P_t^{\mathrm{PV}} \leq P_t^{M,\max}; \quad \forall t \tag{8.27}$$

$$X_{t_0}^b = X_0^b \tag{8.28}$$

$$P_t^{\mathrm{charge}} \leq P_{\mathrm{charge}}^{\max} \times U_t^{\mathrm{charge}}; \quad \forall t \tag{8.29}$$

$$P_t^{\mathrm{disc}} \leq P_{\mathrm{disc}}^{\max} \times U_t^{\mathrm{disc}}; \quad \forall t \tag{8.30}$$

$$X_b^{\min} \leq X_t^b \leq X_b^{\max}; \quad \forall t \tag{8.31}$$

$$U_t^{\mathrm{charge}} + U_t^{\mathrm{disc}} \leq 1; \quad \forall t \tag{8.32}$$

$$X_t^b = X_{t-1}^b + \chi \times P_t^{\mathrm{charge}} - \frac{P_t^{\mathrm{disc}}}{\eta}; \quad \forall t \tag{8.33}$$

$$\mathrm{load}_t = (1 - \mathrm{DR}_t) \cdot \mathrm{load}_t^0 + \mathrm{ldr}_t \tag{8.34}$$

$$\mathrm{load}_t^0 - \mathrm{load}_t = \mathrm{DR}_t \cdot \mathrm{load}_t^0 - \mathrm{ldr}_t \tag{8.35}$$

$$\sum_{t=1}^{T} \mathrm{ldr}_t = \sum_{t=1}^{T} \mathrm{DR}_t \cdot \mathrm{load}_t^0 \tag{8.36}$$

$$\mathrm{load}_t^{\mathrm{inc}} \leq \mathrm{inc}_t \cdot \mathrm{load}_t^0; \quad \forall t \tag{8.37}$$

$$\mathrm{DR}_t \leq \mathrm{DR}_{\max}; \quad \forall t \tag{8.38}$$

$$\mathrm{inc}_t \leq \mathrm{inc}_{\max}; \quad \forall t \tag{8.39}$$

In constraints (8.7)–(8.39), $P_{l,t}^{\mathrm{BC}}$ is the procured power from the BCs with predetermined price $\lambda_{l,t}$; P_t^p is the cost of energy procurement from the pool market with the market price λ_t; $P_{j,h,t}$ is the cost of produced power by the jth unit of the self-generation units; C_t^B is the operating cost of using battery storage, with charging/discharging efficiency χ/η and charging/discharging power of battery storage (MW) $P_t^{\mathrm{charge}}/P_t^{\mathrm{disc}}$; C_t^{wind} is the cost of produced power from the wind turbine; and C_t^{PV} is the cost of produced power P_t^{PV} from the PV system.

Also, load_t^0 is the initial load, $P_{l,t}^{\min}/P_{l,t}^{\max}$ is the minimum/maximum capacity relating, P_t^{BC} is the total procured power from the bilateral contracts, $P_{j,h}^{\mathrm{MAX}}$ is the power output of block h of jth unit of self-generating units, $R_j^{\mathrm{up}}, R_j^{\mathrm{down}}$ is the ramp up/down rate limit of j unit, MUT_j,MDT_j is the minimum up/down time of self-generation units j, V_t^w is the wind speed, V_r is the rated speed, V_{ci}, is the cut-in speed, V_{c0} is the cutout wind speed, $P_t^{M,\max}$ is the maximum power at any temperature T_t^a, G_t^a is irradiation, $P_{\mathrm{Max},0}^M$ is the maximum power, and NOCT is the normal operating cell.

8.5 Simulation Results

The robust energy procurement problem of a large consumer is solved under TOU considering 30% variation in power price in the pool market. Therefore, as said before, the power price in the pool market is considered as uncertain parameter in the robust optimization approach. The problem is formulated as MILP and solved under CPLX solver [16] in the GAMS optimization software [17]. The problem is solved in 11 iterations assigning different values for the robustness parameter in the interval [0, 1], starting from zero to one taking 0.1 step size. In this chapter, first the robust results are presented and discussed in two cases as with and without TOU-DRP; then, the results of the deterministic and worst-case robust optimization are compared in both with and without TOU-DRP cases.

8.5.1 Robust Optimization Results

The power price in the market may be increased or reduced which impose uncertainty in power procurement problem. In this chapter, 30% variation in the power price is assumed which is shown in Fig. 8.2. In the worst case, 30% variation in the power price is increased. Note that robust optimization approach is not considered the best case corresponding to 30% reduction in power price.

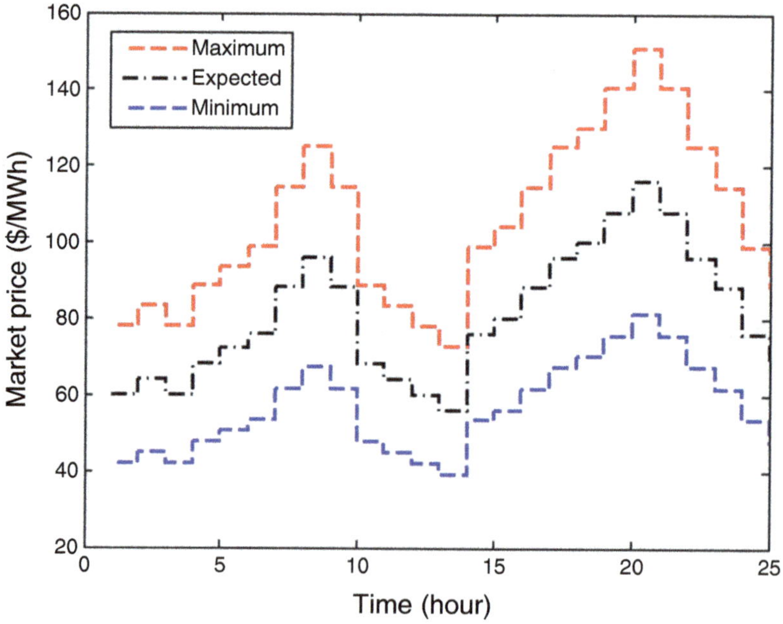

Fig. 8.2 Power price in the pool market under 30% variation

8.5.2 Total Robust Procurement Cost

The total robust procurement cost in both without and with implementing TOU rate of DRP is presented in Fig. 8.3 for different amounts of robustness parameter. It should be noted that assigning zero for the robustness parameter gives deterministic case results. As it can be seen in the implementation, TOU has reduced total power procurement cost in which in the worst case, the total cost is reduced from \$40,949 in without TOU case to \$39,129 in with TOU case indicating about 4.65% reduction. In the determinist case, results of the total power procurement are recorded as \$40,253 and \$38,401 for without TOU and with TOU, respectively. On the other hand, as was expected, by increasing robustness parameter gamma, which presents power price increase in the pool market, the total power procurement cost is increased. Obtained results show that 30% increase in power price resulted 1.79% and 1.86% increase in total cost for without and with TOU cases, respectively. From the abovementioned numbers, it can be interfered that large consumer is not sensitive to power price increase in the pool market which relies on the high capacity of self-generating units, renewable energy sources, and bilateral contracts with predetermined prices. Table 8.1 presents a summary of obtained result of deterministic and robust optimization methods for both with and without TOU cases.

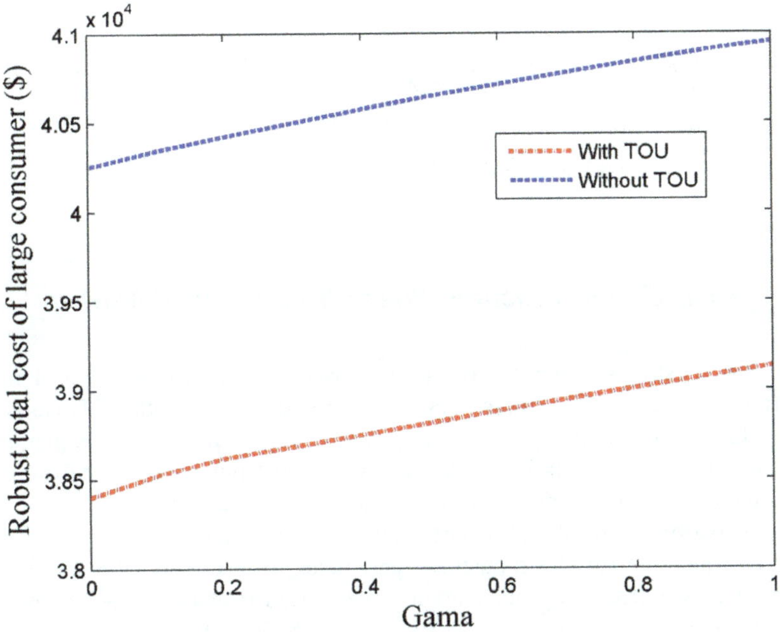

Fig. 8.3 Total robust power procurement cost for with and without TOU

Table 8.1 Summary of obtained results of total power procurement cost

	With TOU	Without TOU
Deterministic	38,401	40,253
Robust	39,129	40,949

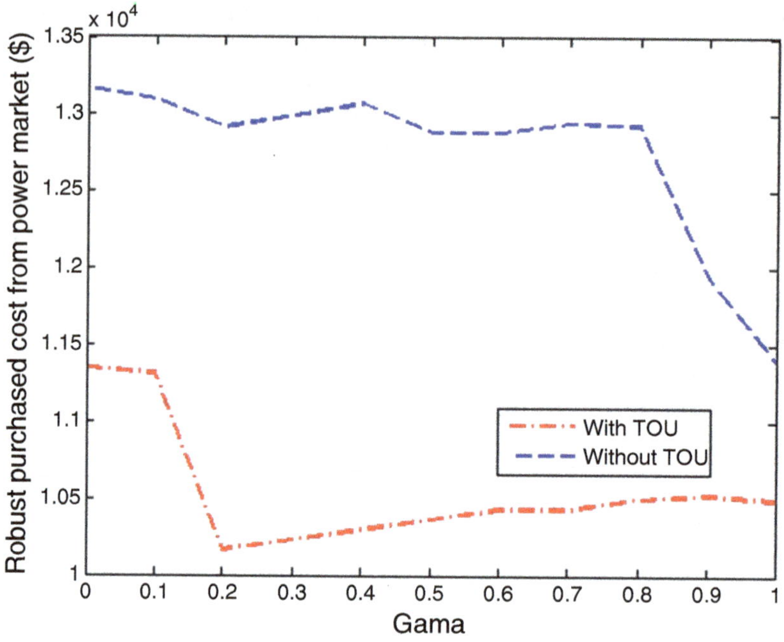

Fig. 8.4 Robust cost of purchased power from the pool market

8.5.3 *Total Cost of Procured Power from the Pool Market*

Figure 8.4 illustrates the total cost of purchased power from the pool market for different values of robustness parameter for without and with TOU cases. It is obvious that robust purchased cost from power market is reduced by implementing TOU rate of DRP. In with TOU case, at first, robust purchased cost from power market increases and then decreases and eventually increases again.

By comparing the results of deterministic and robust cases obtained for considering zero and one for the robustness parameter, respectively, the total cost of purchased power from the pool market is decreased because of increased power prices. Comparing results show 13.45% and 7.53% reduction in total cost of procured power from the pool market for without and with TOU cases. Table 8.2 presents a summary of the obtained results.

Table 8.2 Summary of obtained results of purchased power from the pool market

	Without TOU	With TOU
Deterministic	13,168.363	11,350.531
Robust	11,397.631	10,496.254

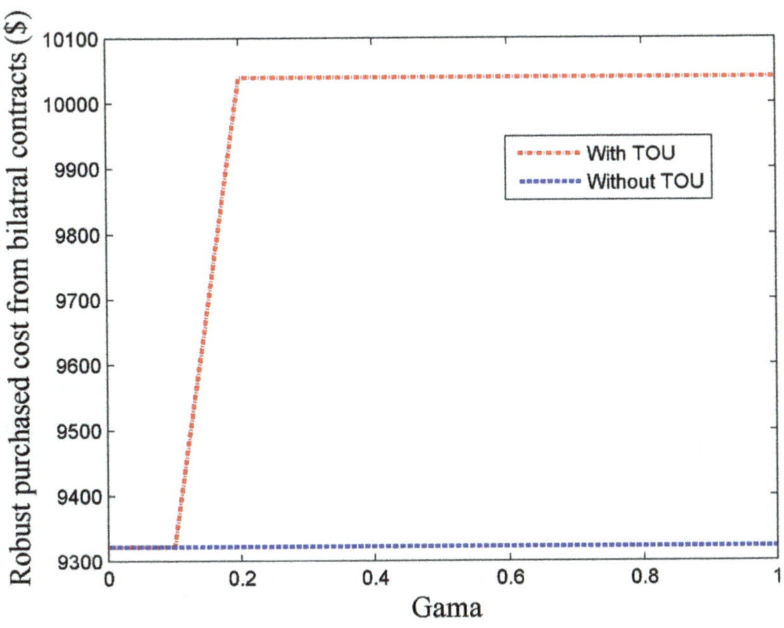

Fig. 8.5 Robust cost of purchased power from the bilateral contracts

8.5.4 Total Cost of Procured Power from the Bilateral Contracts

The robust cost of purchased power from the bilateral contracts is depicted in Fig. 8.5. Purchased power from the bilateral contracts in without implementing TOU is same for different amounts of robustness parameter. In the with TOU case, purchased power from the bilateral contracts is increased by increasing the robustness parameter. This means that in with TOU case, reduced power from the pool market because of increasing power price is compensated by the bilateral contracts. Table 8.3 presents a summary of the obtained results. Based on Table 8.3, the total cost of purchased power from the bilateral contract is increased from $9322.080 to $10,039.964 indicating 7.70% increase due to considering 30% power price increase in the pool market.

Table 8.3 Summary of obtained results of purchased power from the bilateral contracts

	Without TOU	With TOU
Deterministic	9322.080	9322.080
Robust	9322.080	10,039.964

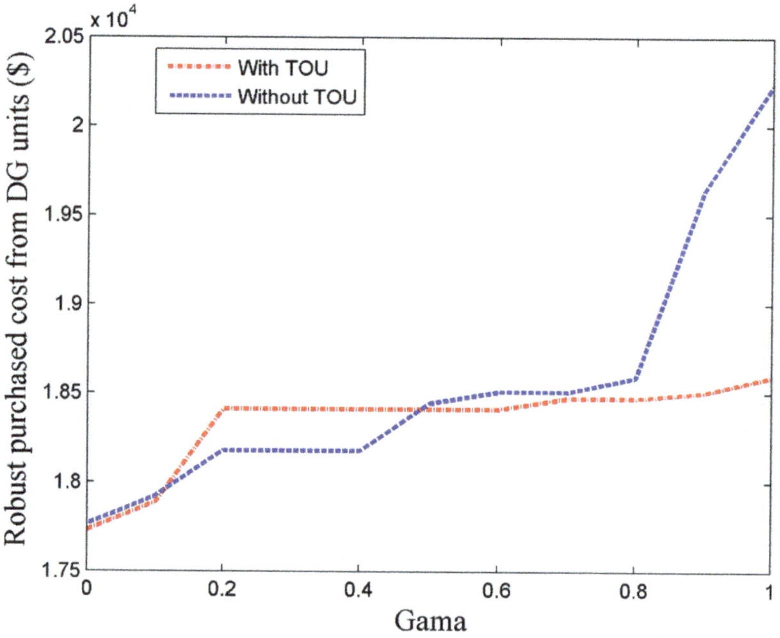

Fig. 8.6 Robust cost of purchased power from the self-generating units

8.5.4.1 Total Cost of Procured Power from the Self-Generating Units

Figure 8.6 shows the procured power from the self-generating units. Most of the required load demand of large consumer is provided by these units. In comparison with deterministic case results, by increasing the robustness parameter, procured power from the self-generating units is increasing. As was expected, the amount of procured power is higher in without TOU case than with TOU, because purchased power from the bilateral contracts is not changed in this case. Obtained results indicate 13.88% and 4.85% increase in operating cost of self-generating units considering the determinist and worst case in robust optimization approach for both without and with TOU cases. A summary of the obtained results for both

Table 8.4 Summary of obtained results of purchased power from the pool market

	Without TOU	With TOU
Deterministic	17,762.808	17,728.427
Robust	20,229.000	18,592.680

without and with TOU cases for both deterministic and robust optimization methods is provided in Table 8.4.

8.6 Comparing Robust and Deterministic Case Results

In the following subsections, procured power from alternative sources are discussed and compared for both deterministic and robust cases under TOU rate of DRP. Note that the results of robust optimization approach are reported for the worst case, considering the robustness parameter equal to one. The result is presented in five different cases as load demand, purchased power from the pool market, self-generating units, bilateral contracts, charging/discharging state, and stored energy in the battery storage system.

8.6.1 Load Profile

The load profile is depicted in Fig. 8.7 for robust and deterministic cases, with and without implementing TOU pricing. The load profile without implementing TOU is same in both robust and deterministic cases. In this situation, maximum load demand is reported at hour 20 equal to 40.5 MW. In addition, the minimum load demand is reported as 11.6 MW at hour 2. After applying TOU rate of DRP, the load profile is changed in both cases regarding power price in the market. Note that in the deterministic case, power price is fixed, while in the robust case, it is increased about 30%. In this condition, the maximum and minimum load demands are recorded as 35.4 and 46.6 MW at hours 13 and 15 for robust and deterministic cases, respectively.

It is obvious that after applying TOU pricing, the load is shifted from high-price periods to low-price periods. It should be reminded that the amount of shifted load is limited on 20% of the base load.

8.6.2 Purchased Power from the Pool Market

Figure 8.8 illustrates the purchased power from the pool market. It can be said that purchased power follows a same pattern in both cases with difference in procured

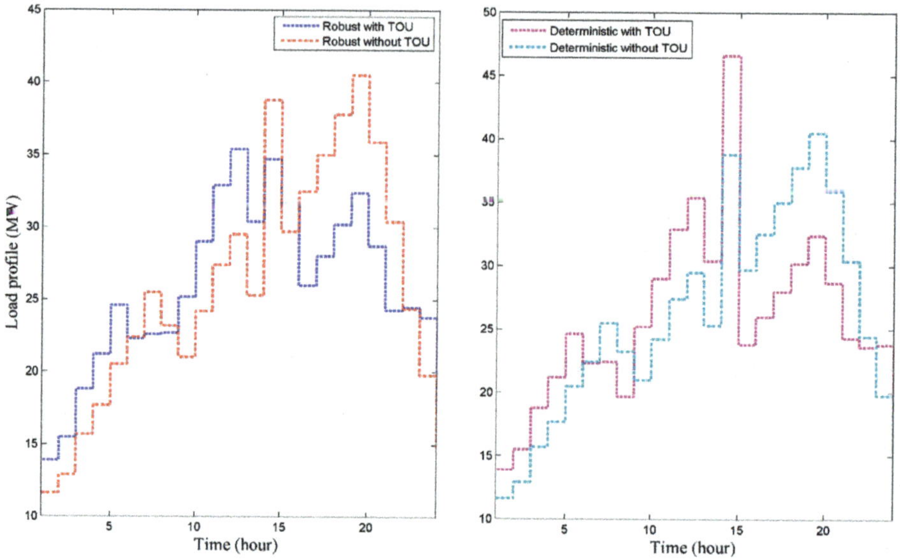

Fig. 8.7 Load profile for deterministic and robust cases under implementing TOU

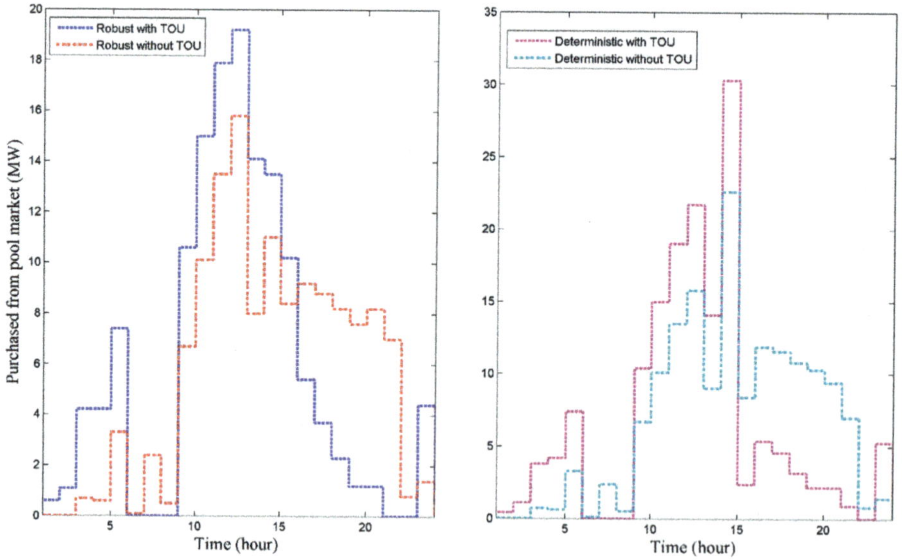

Fig. 8.8 Purchased power from the pool market for deterministic and robust cases under implementing TOU

amount of power, in which purchased power is higher in deterministic case than robust case due to relatively moderate power prices.

Furthermore, after applying TOU rate of DRP, purchased power is reduced in during peak price and load periods both cases in which reduced amount of power is higher in the robust case considering high-power prices in the market.

8.6.3 Procured Power from the Self-Generating Units

Procured power from the self-generating units is presented in Fig. 8.9 for time horizon of the study. It is obvious that there is difference between obtained results for deterministic and robust cases. Without applying TOU, obtained results are different at hours 14–15 and 17–21. The maximum procured power is recorded as 22 MW and 24.7 MW at hour 20 for both deterministic and robust cases, respectively. In addition, the minimum procured power is obtained at 7.3 MW at hour 2 for both cases. After applying TOU-DRP, procured power from the self-generating units is reduced. Comparing the result of these two cases shows that the results are different at hours 8–9, 12–13, and 15. In this case, the maximum procured power is recorded as 22 MW at hour 20 for both deterministic and robust cases.

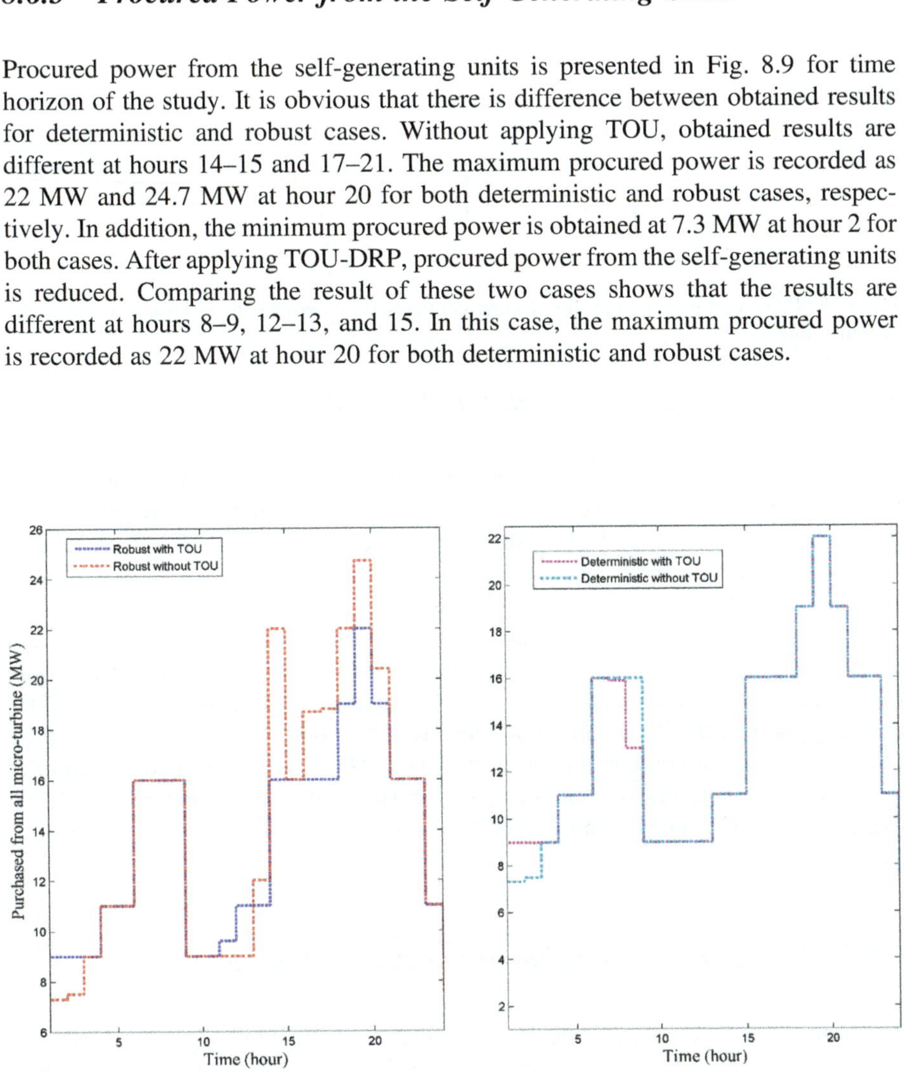

Fig. 8.9 Procured power from the self-generating units for deterministic and robust cases under implementing TOU

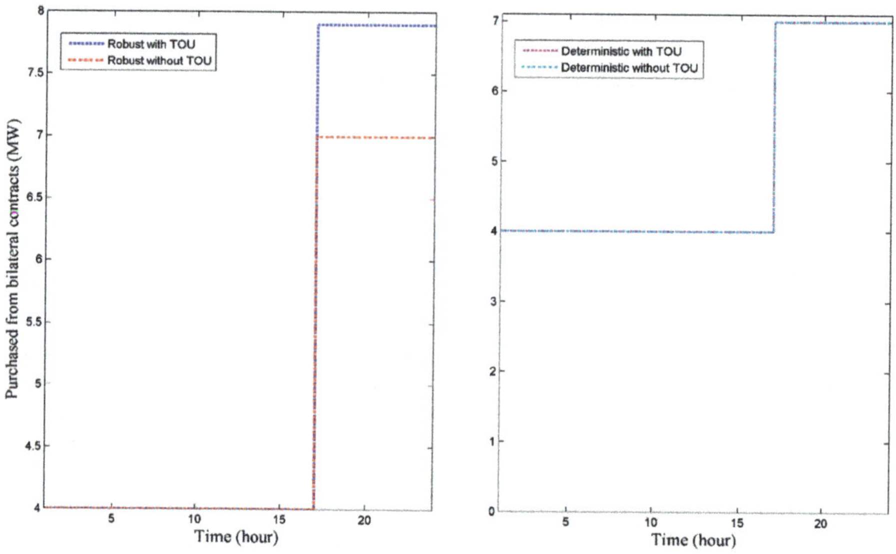

Fig. 8.10 Purchased power from the bilateral contracts for deterministic and robust cases under implementing TOU

8.6.4 Purchased Power from the Bilateral Contracts

Figure 8.10 presents the purchased power from the bilateral contracts. Regarding predetermined prices and periods of bilateral contracts, deterministic results are the same for with and without TOU cases, while after applying TOU in robust optimization, purchased power is increased to compensate reduced procured power from the pool market. The minimum purchased power from the bilateral contracts is reported at hours 2–17 which is equal to 4 MW for both with and without TOU cases in both optimization methods. It should be noted that the required demand of the large consumer is zero at this hour. The maximum procured power for with TOU case is equal to 7.9 MW recorded at hours 18–24 in both deterministic and robust cases. The maximum purchased power in without TOU case is obtained as 7 and 7.9 for robust and deterministic optimizations, respectively. Note that in Fig. 8.10, aggregation of 12 bilateral contracts is depicted at each period.

8.6.5 Charging/Discharging of the Battery Storage System

The charging or discharging state of the battery is illustrated in Fig. 8.11. Obtained results for without TOU case, in both deterministic and robust optimization methods, only differ at hours 14, 19, and 21, while after applying TOU, differences are seen at hours 2, 4, 10, and 12–13. Note that due to small capacity of the battery storage

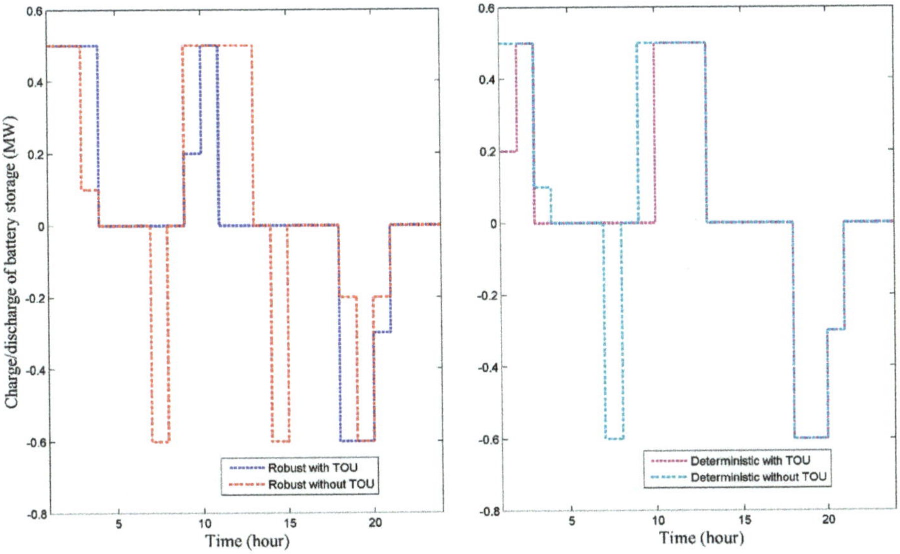

Fig. 8.11 Charging/discharging of the battery storage system for deterministic and robust cases under implementing TOU

system in the considered case study, its impact on power procurement process is negligible in which maximum charged/discharged power is equal to 0.6 MW in both cases of the optimization methods.

8.6.6 Stored Energy in the Battery Storage System

Finally, Fig. 8.12 presents the stored energy in the battery storage system for 24 h. In without TOU case, the total stored energy in the battery is reported as 25.5 and 22.3 MWh for the deterministic and robust optimization methods, respectively.

After applying TOU rate of DRP, the number is changed to 24.7 and 30.7 MWh for the deterministic and robust optimization methods, respectively.

The minimum and maximum amount of stored energy in the battery storage system is equal to 0.2 and 2 MW for both with and without TOU cases in both deterministic and robust optimization methods.

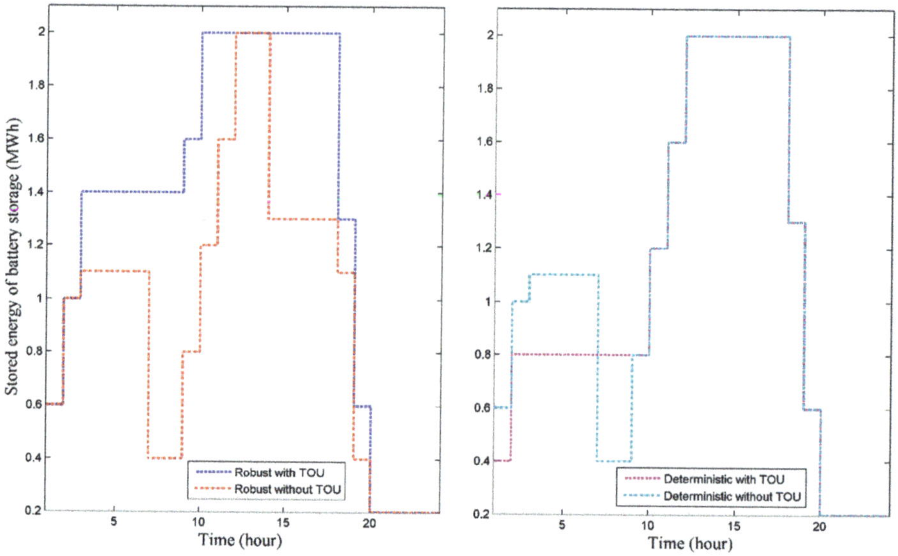

Fig. 8.12 Stored energy in the battery storage system for deterministic and robust cases under implementing TOU

8.7 Conclusion

In this chapter, the impact of TOU rate of DRP on robust energy procurement of a large consumer is investigated. The problem is solved considering 30% increase in power price in the pool market using the robust optimization approach. The obtained results are presented for different values of the robustness parameter and compared with deterministic case results under two cases as without and with TOU.

The total robust power procurement cost is reduced from $40,949 to $39,129 after implementing TOU rate of DRP indicating about 4.65% reduction, while in determinist case, the total power procurement is recorded as $40,253 and $38,401 for without TOU and with TOU, respectively. Obtained results show that 30% increase in power price resulted in 1.79% and 1.86% increase in total cost for without and with TOU cases, respectively.

The total cost of purchased power from the pool market is obtained as $11,397.631 and $10,496.254 for without and with TOU in robust optimization approach. Comparing results of deterministic and robust optimization methods show 13.45% and 7.53% reduction in total cost of procured power from the pool market for without and with TOU cases.

The total cost of purchased power from the bilateral contract is reported as $9322.080 and $10,039.964 for without and with TOU cases in the robust optimization approach. In the deterministic case, the total cost is equal to $9322.080 in both cases. In the robust optimization, 7.70% increase is recorded in total cost due to considering 30% power price increase in the pool market after applying TOU.

The total operating cost of self-generating units are reported as \$20,229.000 and \$18,592.680 for without and with TOU cases in the robust optimization approach. The results of deterministic optimization are obtained as \$17,762.808 and \$17,728.427 for without and with TOU cases. Obtained results indicate 13.88% and 4.85% increase in operating cost of self-generating units considering the determinist and worst case in robust optimization approach for both without and with TOU cases.

Finally, the procured power from each of the alternative sources is presented and compared for robust and deterministic optimization methods in without TOU and with TOU cases.

References

1. Y. Wang, L. Li, Time-of-use electricity pricing for industrial customers: a survey of U.S. utilities. Appl. Energy **149**, 89–103 (2015)
2. N. Ghadimi, Two new methods for power flow tracing using bus power balance equations. J. Cent. South Univ. **21**(7), 2712–2718 (2014)
3. U.S. Energy Information Administration, *Smart Grid Legislative and Regulatory Policies and Case Studies* (Author, Washington, DC, 2014)
4. M. Hosseini Firouz, N. Ghadimi, Optimal preventive maintenance policy for electric power distribution systems based on the fuzzy AHP methods. Complexity **21**(6), 70–88 (2016)
5. M.P. Lee, Assessment of Demand Response & Advanced Metering, Staff Report (2012)
6. H.A. Aalami, M.P. Moghaddam, G.R. Yousefi, Modeling and prioritizing demand response programs in power markets. Electr. Power Syst. Res. **80**(4), 426–435 (2010)
7. S. Nojavan, H. allah Aalami, Stochastic energy procurement of large electricity consumer considering photovoltaic, wind-turbine, micro-turbines, energy storage system in the presence of demand response program. Energy Convers. Manag. **103**, 1008–1018 (2015)
8. S. Nojavan, H. Ghesmati, K. Zare, Robust optimal offering strategy of large consumer using IGDT considering demand response programs. Electr. Power Syst. Res. **130**, 46–58 (2016)
9. S. Nojavan, K. Zare, B. Mohammadi-Ivatloo, Robust bidding and offering strategies of electricity retailer under multi-tariff pricing. Energy Econ. **68**, 359–372 (2017)
10. S. Nojavan, K. Zare, B. Mohammadi-Ivatloo, Selling price determination by electricity retailer in the smart grid under demand side management in the presence of the electrolyser and fuel cell as hydrogen storage system. Int. J. Hydrog. Energy **42**(5), 3294–3308 (2017)
11. S. Fan, R.J. Hyndman, The price elasticity of electricity demand in South Australia. Energy Policy **39**(6), 3709–3719 (2011)
12. A. Faruqui, S. Sergici, Household response to dynamic pricing of electricity: a survey of 15 experiments. J. Regul. Econ. **38**(2), 193–225 (2010)
13. A. Faruqui, S. Sergici, N. Lessem, D. Mountain, F. Denton, B. Spencer, C. King, *Impact Evaluation of Ontario's Time-of- Use Rates: First Year Analysis* (The Brattle Group, Cambridge, 2013)
14. L. Ji, D.X. Niu, G.H. Huang, An inexact two-stage stochastic robust programming for residential micro-grid management-based on random demand. Energy **67**, 186–199 (2014)
15. S. Nojavan, H. Qesmati, K. Zare, H. Seyyedi, Large consumer electricity acquisition considering time-of-use rates demand response programs. Arab. J. Sci. Eng. **39**(12), 8913–8923 (2014)
16. CPLEX 12. [Online]. Available: https://www.gams.com/latest/docs/S_CPLEX.html. Accessed 15 Jul 2018
17. A. Brooke, D. Kendrick, A. Meeraus, R. Raman, R.E. Rosenthal, *GAMS a User's Guide Introduction 1* (GAMS Development Corporation, Washington, DC, 1998)

Chapter 9
Robust Energy Procurement Under Real-Time Pricing

Alireza Rezvani

Abstract In order to reduce power consumption and operating cost of power generation units, demand response programs can be considered as an effective and essential tool. As a key component of demand response, real-time pricing (RTP) encourages consumers in an economical and efficient way. In this chapter, RTP scheme is implemented to manage peak load for the power procurement problem of a large consumer. Some part of the required power of the large consumer procured from the pool market and power price in the market has uncertainty. So, the robust optimization method is used to model the uncertainty in the system.

The results of using RTP are presented for a case study, and obtained results are analyzed and compared with deterministic and time-of-use demand response program which are detained in Chaps. 4 and 8, respectively.

9.1 Introduction

The importance of implementing demand response programs (DRPs) is detailed in the previous chapter. In this chapter, the impact of implementing real-time price (RTP) on robust energy procurement problem faced by a large consumer is investigated. Real-time pricing rate of demand response programs is an effective strategy to reduce electricity consumption cost, which is categorized as time-based program (TBP).

The impact of RTP demand response program on Chinese household's 24-h electricity demand profile is investigated in [1] based on household survey and agent-based modeling. Obtained results indicate that RTP has the potential to shift electricity load from peak to off-peak load periods and reduce the total electricity consumption [2, 3].

A. Rezvani (✉)
Young Researchers and Elite Club, Saveh Branch, Islamic Azad University, Saveh, Iran

Iran Water and Power Resources Development Company (IWPCO), Tehran, Iran

S. Nojavan et al. (eds.), *Robust Energy Procurement of Large Electricity Consumers*,
https://doi.org/10.1007/978-3-030-03229-6_9

The RTP is implemented on an agent-based model combining both spot and balancing electricity markets in [4]. The results show that RTP prices may lead to higher electricity procurement costs, which contradicts with the usual claim that RTP scheme reduces the electricity procurement cost which is a systemic defect.

A new framework is proposed in [5] to generate feasible RTP curves for end-user customers in a demand response management process. In order to show the impact of the derived curves, a case simulation study of the Italian market is provided.

For industrial customers, the pricing components and characteristics of representative RTP programs are analyzed in [6] based on available information. In this Ref., the success and failure reasons of RTP for large industrial consumer are studied in US electricity market.

In this chapter, the considered case study in Chap. 2 is used to analyze the impact of RTP rate of DRP. Obtained results are presented for two cases as with and without RTP. In addition, the results are compared with deterministic case. Figure 9.1 presents the configuration of the case study.

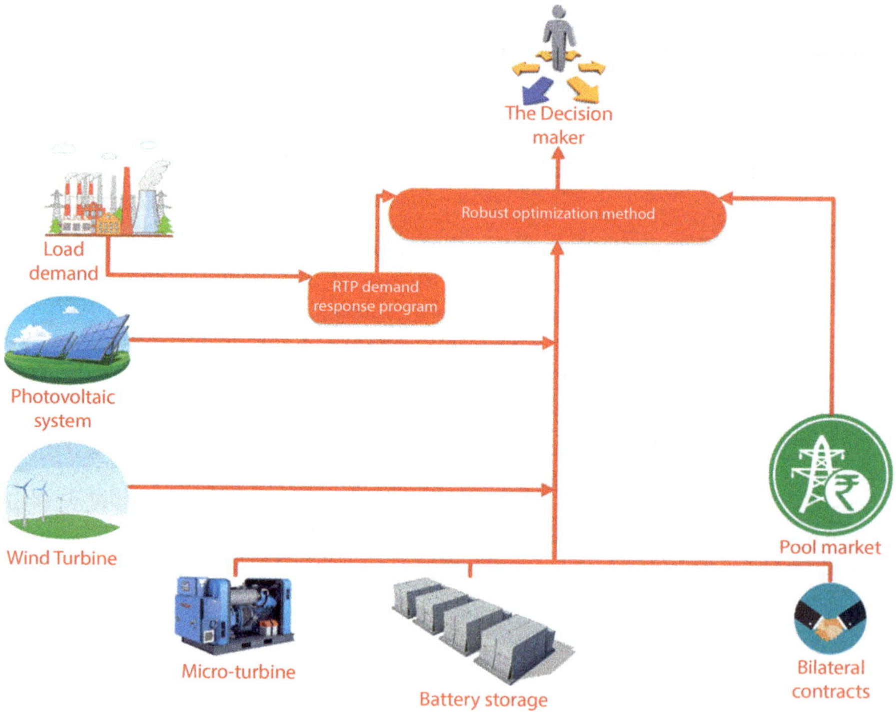

Fig. 9.1 Configuration of case study

9.2 Modeling of RTP

In this section, formulation of RTP is developed to solve the power procurement problem of a large consumer. The RTP model is developed based on predicted information of electricity demand [7]. Therefore, the total demand of a large consumer is presented and should be calculated as shown in Eq. (9.1).

$$T_d = \sum_{t=1}^{24} P_t^D \tag{9.1}$$

where

T_d is the total load demand of large consumer
P_t^D is the initial electric load demand at tth load level in MWh

In order to get the float factor of RTP, first the average electricity demand should be calculated as presented in Eq. (9.2) [4]. Then, Eq. (9.3) is used to calculate the float factor of RTP γ_t.

$$P_{\text{av}} = \frac{T_d}{24} \tag{9.2}$$

$$\gamma_t = \frac{P_t^D}{P_{\text{av}}} \tag{9.3}$$

Finally, using Eqs. (9.4) and (9.5), the RTP model is obtained.

$$\lambda_{\text{RTP}} = \gamma_t \cdot \lambda_{\text{TOU}} \tag{9.4}$$

$$\lambda_{\text{RTP}}^{\text{Min}} \leq \lambda_{\text{RTP}} \leq \lambda_{\text{RTP}}^{\text{Max}} \tag{9.5}$$

where

λ_{TOU} is the benchmark price which is the time-of-use pricing
$\lambda_{\text{RTP}}^{\text{Min}}$ is the minimum limit of the RTP
$\lambda_{\text{RTP}}^{\text{Max}}$ is the maximum limit of the RTP

Based on the abovementioned formulation, the new load profile under RTP rate of DRP is calculated using Eq. (9.6) [8].

$$P_t^{\text{DR}} = P_t^D + \frac{E \cdot P_t^D(\lambda_{\text{RTP}} - \lambda_{\text{TOU}})}{\lambda_{\text{RTP}}} \tag{9.6}$$

where the E is the demand price elasticity coefficient. It should be denoted that E is determined by analyzing the customer types and historical demand data. According to [9], E can take values in the interval [−0.5, 0]. In this chapter, E is considered equal to −0.5.

9.3 Formulation of Robust Energy Procurement

In order to solve robust energy procurement problem of a large consumer under RTP rate of DRP, the robust optimization method is used in this chapter. Detailed information about the robust optimization approach is provided in Chap. 4. Therefore, in this chapter, the formulation of robust energy procurement of large consumer is briefly introduced under the RTP rate of DRP.

$$\text{Min} \quad \text{OF} = z \cdot \Gamma + \sum_{t=1}^{24} q_t + \sum_{l=1}^{B} \sum_{t=1}^{T} \lambda_{l,t} P_{l,t}^{\text{BC}} + \sum_{t=1}^{T} \left\{ \begin{array}{l} \lambda_t P_t^p + \sum_{j=1}^{N_j} \sum_{h=1}^{N_h} S_{j,h}^{\text{DG}} P_{j,h,t} \\ + C_t^B \times \left(\chi \times P_t^{\text{charge}} + \dfrac{P_t^{\text{disc}}}{\eta} \right) \\ + C_t^{\text{wind}} \times P_t^{\text{wind}} + C_t^{\text{PV}} \times P_t^{\text{PV}} \end{array} \right\} \tag{9.7}$$

Subject to

$$z + q_t \geq \widehat{\lambda}_t \cdot y_t \forall t \mid \widehat{\lambda}_t > 0 \tag{9.8}$$

$$q_t \geq 0 \forall t \mid \widehat{\lambda}_t > 0 \tag{9.9}$$

$$y_t \geq 0 \tag{9.10}$$

$$z \geq 0 \tag{9.11}$$

$$P_t^p \leq y_t \tag{9.12}$$

$$q_t \geq 0 \forall t \mid \widehat{\lambda}_t > 0 \tag{9.13}$$

$$\sum_{l=1}^{B} P_{l,t}^{\text{BC}} + P_t^p + \sum_{j=1}^{N_j} \sum_{h=1}^{N_h} P_{j,h,t} + P_t^{\text{wind}} + P_t^{\text{PV}} + P_t^{\text{disc}} = P_t^{\text{DR}} + P_t^{\text{charge}} \tag{9.14}$$

$$P_{l,t}^{\min} s_l < P_{l,t}^{\text{BC}} < P_{l,t}^{\max} s_l; \quad \forall l, t \tag{9.15}$$

$$P_t^{\text{BC}} = \sum_{l=1}^{N_l} P_{l,t}^{\text{BC}}; \quad \forall t \tag{9.16}$$

$$\text{Cost} = \sum_{t=1}^{T} \sum_{j=1}^{N_j} \sum_{h=1}^{N_h} S_{j,h}^{\text{DG}} P_{j,h,t} \tag{9.17}$$

$$0 \leq P_{j,h,t} \leq \left(P_{j,h}^{\mathrm{MAX}} - P_{j,h-1}^{\mathrm{MAX}}\right) \times U_{j,t}; \quad \forall j,h,t \tag{9.18}$$

$$0 \leq P_{j,1,t} \leq P_{j,1}^{\mathrm{MAX}} \times U_{j,t}; \quad \forall j,t \tag{9.19}$$

$$\sum_{h=1}^{N_h} P_{j,h,t} - \sum_{h=1}^{N_h} P_{j,h,t-1} \leq R_j^{\mathrm{up}} \times U_{j,t}; \quad \forall j,t \tag{9.20}$$

$$\sum_{h=1}^{N_h} P_{j,h,t-1} - \sum_{h=1}^{N_h} P_{j,h,t} \leq R_j^{\mathrm{down}} \times U_{j,t-1}; \quad \forall j,t \tag{9.21}$$

$$\begin{array}{l} U_{j,t} - U_{j,t-1} \leq U_{j,t+\mathrm{Up}_{j,i}}; \quad \forall j, \forall t, \forall i \\ U_{j,t-1} - U_{j,t} \leq 1 - U_{j,t+\mathrm{Dn}_{j,i}}; \quad \forall j, \forall t, \forall i \end{array} \tag{9.22}$$

$$\begin{array}{l} \mathrm{Up}_{j,i} = \left\{ \begin{array}{ll} i & i \leq \mathrm{MUT}_j \\ 0 & i \succ \mathrm{MUT}_j \end{array} \right\} \\ \mathrm{Dn}_{j,i} = \left\{ \begin{array}{ll} i & i \leq \mathrm{MDT}_j \\ 0 & i \succ \mathrm{MDT}_j \end{array} \right\} \end{array} \tag{9.23}$$

$$P_{t,s}^{\mathrm{wind,\,max}} = \begin{cases} 0 & V_{t,s}^{w} < V_{\mathrm{ci}} \\ p_r \times \left(\frac{V_{t,s}^{w} - V_{\mathrm{ci}}}{V_r - V_{\mathrm{ci}}}\right)^3 & V_{\mathrm{ci}} < V_{t,s}^{w} < V_{\mathrm{cr}} \\ p_r & V_r < V_{t,s}^{w} < V_{c0} \\ 0 & V_{t,s}^{w} > V_{c0} \end{cases} \tag{9.24}$$

$$P_t^{\mathrm{wind}} \leq P_t^{\mathrm{wind,\,max}}; \quad \forall t \tag{9.25}$$

$$P_t^{M,\,\max} = \frac{G_t^a}{G_{a_0}} \times \left\{ P_{\mathrm{Max},0}^{M} + \mu_{P\max} \times \left(T_{t,}^{a} + G_t^a \times \frac{\mathrm{NOCT} - 20}{800} - T_{M,0} \right) \right\} \tag{9.26}$$

$$P_t^{\mathrm{PV}} \leq P_t^{M,\,\max}; \quad \forall t \tag{9.27}$$

$$X_{t_0}^{b} = X_0^{b} \tag{9.28}$$

$$P_t^{\mathrm{charge}} \leq P_{\mathrm{charge}}^{\max} \times U_t^{\mathrm{charge}}; \quad \forall t \tag{9.29}$$

$$P_t^{\mathrm{disc}} \leq P_{\mathrm{disc}}^{\max} \times U_t^{\mathrm{disc}}; \quad \forall t \tag{9.30}$$

$$X_b^{\min} \leq X_t^{b} \leq X_b^{\max}; \quad \forall t \tag{9.31}$$

$$U_t^{\mathrm{charge}} + U_t^{\mathrm{disc}} \leq 1; \quad \forall t \tag{9.32}$$

$$X_t^{b} = X_{t-1}^{b} + \chi \times P_t^{\mathrm{charge}} - \frac{P_t^{\mathrm{disc}}}{\eta}; \quad \forall t \tag{9.33}$$

$$T_d = \sum_{h=1}^{24} P_t^{D} \tag{9.34}$$

$$P_{\text{av}} = \frac{T_d}{24} \tag{9.35}$$

$$\gamma_t = \frac{P_t^D}{P_{\text{av}}} \tag{9.36}$$

$$\lambda_{\text{RTP}} = \gamma_t \cdot \lambda_{\text{TOU}} \tag{9.37}$$

$$\lambda_{\text{RTP}}^{\text{Min}} \leq \lambda_{\text{RTP}} \leq \lambda_{\text{RTP}}^{\text{Max}} \tag{9.38}$$

$$P_t^{\text{DR}} = P_t^D + \frac{E \cdot P_t^D (\lambda_{\text{RTP}} - \lambda_{\text{TOU}})}{\lambda_{\text{RTP}}} \tag{9.39}$$

In constraints (9.7)–(9.39), $P_{l,t}^{\text{BC}}$ is the procured power from the BCs with predetermined price $\lambda_{l,t}$; P_t^p is the cost of energy procurement from the pool market with the market price λ_t; $P_{j,h,t}$ is the cost of produced power by the jth unit of the self-generation units; C_t^B is the operating cost of using battery storage, with charging/discharging efficiency χ/η and charging/discharging power of battery storage (MW) $P_t^{\text{charge}}/P_t^{\text{disc}}$; C_t^{wind} is the cost of produced power from the wind turbine; and C_t^{PV} is the cost of produced power P_t^{PV} from the PV system.

Also, $P_{l,t}^{\min}/P_{l,t}^{\max}$ is the minimum/maximum capacity relation, P_t^{BC} is the total procured power from the bilateral contracts, $P_{j,h}^{\text{MAX}}$ is the power output of block h of jth unit of self-generating units, $R_j^{\text{up}}, R_j^{\text{down}}$ is ramp up/down rate limit of j unit, MUT_j,MDT_j is the minimum up/down time of self-generation units j, V_t^w is the wind speed, V_r is the rated speed, V_{ci}, is the cut-in speed, V_{c0}, is the cutout wind speed, $P_t^{M,\max}$ is the maximum power at any temperature T_t^a, G_t^a is the irradiation, $P_{\text{Max},0}^M$ is the maximum power, and NOCT is the normal operating cell.

In the abovementioned equations, constraints (9.8)–(9.13) are imposed by the robust optimization method. Constraint (9.14) presents the power balance between procured power from alternative sources and required load demand. Related constraints to bilateral contracts are provided by constraints (9.15) and (9.16). The self-generating units are modeled by constraints (9.17)–(9.23). The wind turbine is modeled using constraints (9.24) and (9.25) [10]. Constraints (9.26) and (9.27) model the photovoltaic systems [11]. The model of the battery storage system is provided by constraints (9.28)–(9.33). Finally, RTP rate of DRP is presented by constraints (9.34)–(9.39).

9.4 Simulation Results

In a used manner in Chap. 8, the robust energy procurement problem of a large consumer under TOU rate of DRP is formulated as mixed integer linear programming method and solved using the CPLEX solver in the GAMS optimization software considering 30% variation in power price in the pool market.

In this section, the obtained results for power procurement are reported under implementing RTP rate of DRP. Later the results are compared with deterministic and without DRP cases.

9.4.1 Total Robust Procurement Cost

Figure 9.2 presents the robust procurement cost for different values of the robustness parameter. The results are obtained by solving (9.7)–(9.39). In the worst case, total procurement cost is obtained as $41,235.755 without implementing RTP. After applying RTP-DRP, this number is reduced to $40,526.599, which shows 1.72% reduction in total robust power procurement cost. Note that in the worst-case analysis, 30% increase in power price in the market is considered. In the deterministic case, without assuming any increase in power price, total power price is recorded as $40,585.453 and $39,692.237 for without and with RTP cases, respectively, indicating 2.2% reduction. It is obvious that by increasing power price in the market, the total power procurement cost is increased in which in without RTP case it is increased from $40,585.453 to $41,235.755 and from $39,692.237 to $40,526.599 in with RTP case indicating 1.6% and 2.1% increase in total procurement cost for without and with RTP cases, respectively.

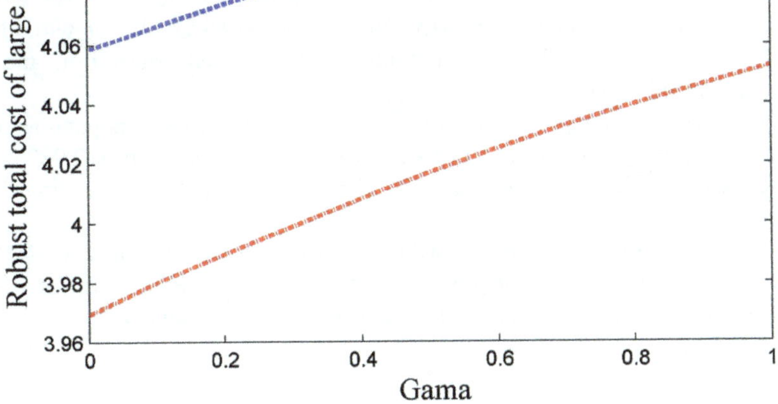

Fig. 9.2 Total robust power procurement cost for with and without RTP

Table 9.1 Summary of obtained results of total power procurement cost

	With RTP	With TOU	Without DRP
Deterministic	40,585.453	38,401	40,585.453
Robust	39,692.237	39,129	41,235.755
Reduction of total cost	1.72%	4.65%	–

Based on obtained results, it can be construed that the large consumer is highly robust against the power price uncertainty in the pool market. For example, as said before, in with RTP-DRP case, the total power procurement cost just increased 2.2% for 30% increase in power price in the pool market because of the high capacity of self-generating units, signing bilateral contracts with predetermined price periods, using renewable energy sources and energy storage systems.

Results of implementing time-of-use rate of DRP, which is detailed in Chap. 8, show 4.65% reduction in the worst-case simulation results meaning in this particular problem better results are obtained using the time-of-use rate of DRP. A summary of results are provided in Table 9.1.

9.4.2 Total Cost of Procured Power from the Pool Market

As it can be predicted, procured power from the pool market is reduced due to power price uncertainty. Total cost of procured power from the pool market versus robust parameter is shown in Fig. 9.3. As was expected, by increasing power price in the market, procured power from the pool market is reduced, and the required load of the consumer is supplied by alternative sources.

According to Fig. 9.3, in the deterministic case results, total cost of procured power from the pool market is recorded as $15,430.565 and $12,932.037 for without and with RTP, respectively, indicating that 16.19% reduction is obtained from procured power because of implementing RTP. In the worst-case results, the cost of procured energy is obtained as $13,239.416 and $12,153.746 for without and with RTP. It is seen that by implementing RTP, procured power from the pool market is reduced about 8.2%.

By increasing power price, which is modeled by the robustness parameter, the cost of purchased power from the pool market is reduced from $12,932.037 to $12,153.746 in with RTP case and from $15,430.565 to $13,239.416 in without RTP case.

Under time-of-use rate of DRP, based on the worst case, results show 13.45% and 7.53% reduction in total cost of procured power from the pool market for without and with TOU cases. A summary of results is presented in Table 9.2.

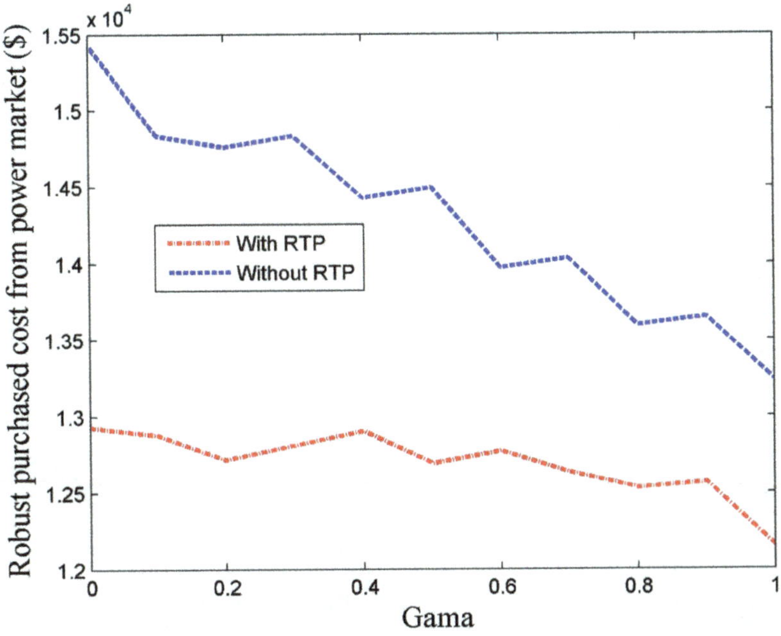

Fig. 9.3 Robust cost of purchased power from the pool market

Table 9.2 Summary of obtained results of purchased power from the pool market

	With RTP	With TOU	Without DRP
Deterministic	12,932.037	11,350.531	15,430.565
Robust	12,153.746	10,496.254	13,239.416
Reduction of total cost	6%	7.53%	–

9.4.3 Total Cost of Procured Power from the Bilateral Contracts

Total cost of purchased power from the bilateral contracts is depicted in Fig. 9.4. In this case, purchased power from the bilateral contracts is not changed by increasing power price in the market. For with and without RTP cases, the total cost is obtained as $9322.080.

From Table 9.3, by applying time-of-use rate, purchased power from the bilateral contracts is increased to $10,039.964.

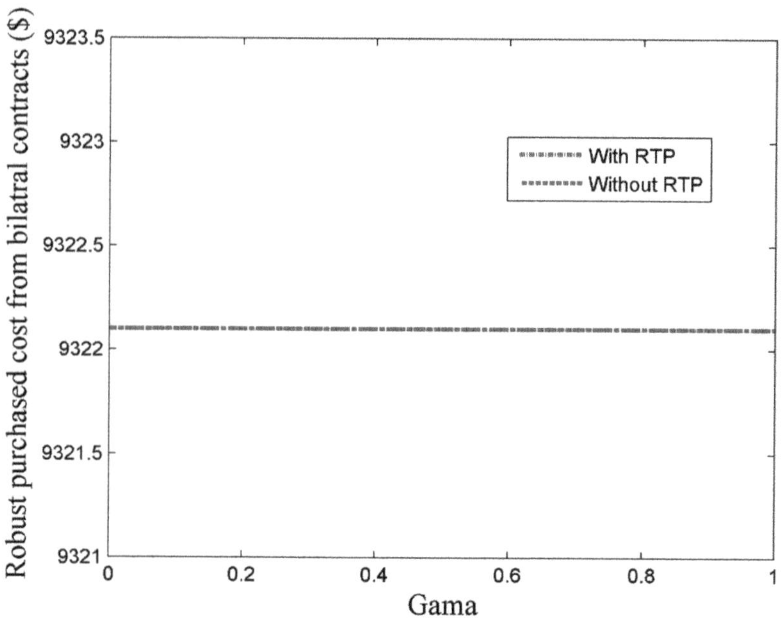

Fig. 9.4 Robust cost of purchased power from the bilateral contracts

Table 9.3 Summary of obtained results of purchased power from the bilateral contracts

	With RTP	With TOU	Without DRP
Deterministic	9322.080	9322.080	9322.080
Robust	9322.080	10,039.964	9322.080

9.4.4 *Total Cost of Procured Power from the Self-Generating Units*

Figure 9.5 shows the procured power from the self-generating units. Note that the self-generating units are the main power procurement source of the required load demand of the large consumer. According to Fig. 9.5, the total operating cost of self-generating units is continuously increasing in both with and without RTP cases. Reduced procured power from the pool market due to high power price is compensated with purchased power from the self-generating units. In determinist case, total operating cost of self-generating units which is considered as procurement cost is recorded as $15,832.808 and $17,438.120 for without and with RTP, respectively. It is obvious that in with RTP case, the operating cost is higher than without RTP case because more power is required for both cases. In the worst case, results are obtained as $18,674.259 and $19,050.773 for without and with RTP cases, respectively. A summary of the obtained results is provided in Table 9.4.

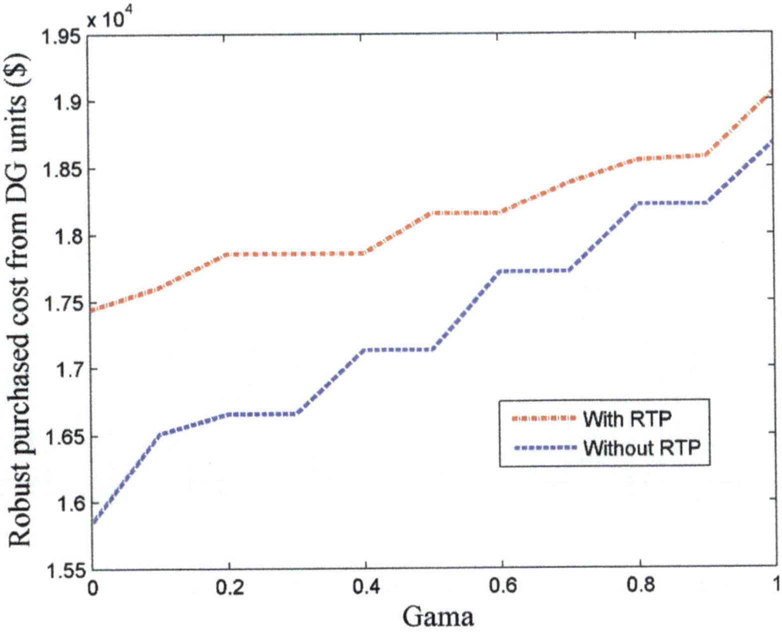

Fig. 9.5 Robust cost of purchased power from the self-generating units

Table 9.4 Summary of obtained results of purchased power from the self-generating units

	With RTP	Without TOU	With DRP
Deterministic	15,832.808	17,438.120	17,728.427
Robust	18,674.259	19,050.773	18,592.680

9.5 Comparing Robust and Deterministic Case Results Under RTP-DRP

In the following subsections, procured power from alternative sources is discussed and compared for both deterministic and robust cases under RTP rate of DRP.

The results are presented in five different cases as load demand, purchased power from the pool market, self-generating units, bilateral contracts, charging/discharging state, and stored energy in the battery storage system.

9.5.1 Load Profile

Figure 9.6 illustrates the load profile for with and without implementing RTP for both deterministic and robust optimization methods. Note that results of robust

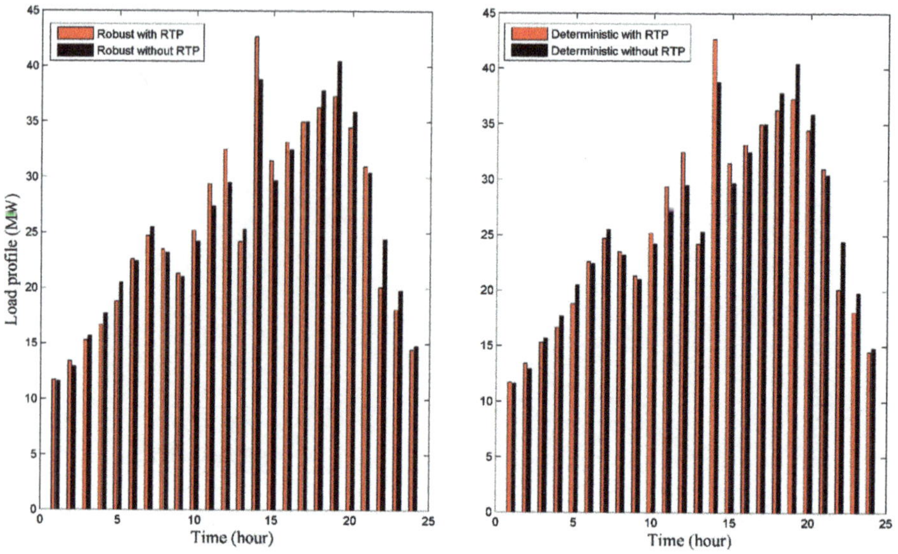

Fig. 9.6 Load profile for deterministic and robust cases under implementing RTP

optimization method are reported for the worst case. In without implementing RTP, the maximum load demand is reported at hour 20 equal to 40.5 MW. In addition, the minimum load demand is reported as 11.6 MW at hour 1. Real-time pricing rate of DRP has changed the load profile considering power price in the market. Under RTP rate of DRP, the maximum and minimum load demands are recorded as 42.7 and 11.7 MW at hours 15 and 1 for both robust and deterministic cases. It is obvious that same results are obtained after applying RTP-DRP.

9.5.2 *Purchased Power from the Pool Market*

The purchased power from the pool market is shown in Fig. 9.7 for time horizon of the study. It can be said that purchased power follows the same pattern in both cases with difference in procured amount of power, in which purchased power is higher in deterministic case than robust case due to relatively moderate power prices.

In the determinist optimization without implementing RTP, the maximum and minimum purchased powers from the pool market are recorded as 17.2 and 0.1 MW at hours 14 and 6, respectively. Under the RTP-DRP, the maximum and minimum purchased powers are obtained as 26.2 and 0.4 MW at hours 14 and 6, respectively.

Results of robust optimization without RTP, 13.9 and 0.1 MW at hours 12 and 6, are reported as the maximum and minimum purchased powers, respectively. Furthermore, under with RTP, the results are obtained as the maximum and minimum purchased powers obtained as 19.7 and 0.4 MW at hours 12 and 6, respectively.

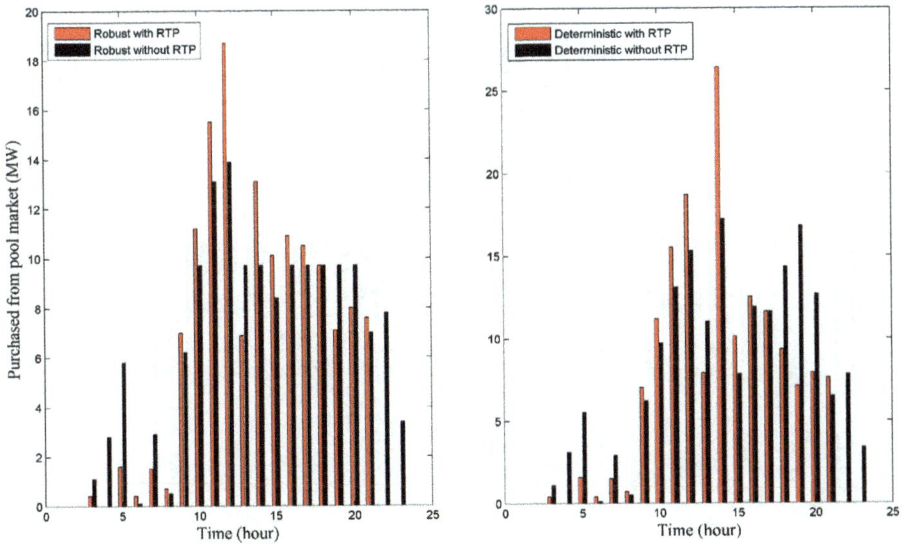

Fig. 9.7 Procured power from the pool market for deterministic and robust cases under implementing RTP

Total purchased powers from the power market in deterministic case are at 178.5 and 157.4 MWh for with and without RTP case. In the robust case, obtained results are 150.60 and 140.90 MWh.

9.5.3 Purchased Power from the Bilateral Contracts

Figure 9.8 presents the purchased power from the bilateral contracts. As discussed before, purchased power from the bilateral contracts are same for robust and deterministic optimizations. The minimum purchased power from the bilateral contracts is reported at hours 1–16 which is equal to 4 MW for both optimization methods in with and without RTP cases. In addition, the maximum procured power is equal to 7 MW recorded at hours 17–24 in both deterministic and robust optimization methods. Note that in Fig. 9.8, aggregation of 12 bilateral contracts is depicted at each period.

9.5.4 Procured Power from the Self-Generating Units

Figure 9.9 presents the generated power of self-generating units at each period for both determinist and robust optimization methods. It should be reminded that the self-generating units supply the significant part of required load demand of the large

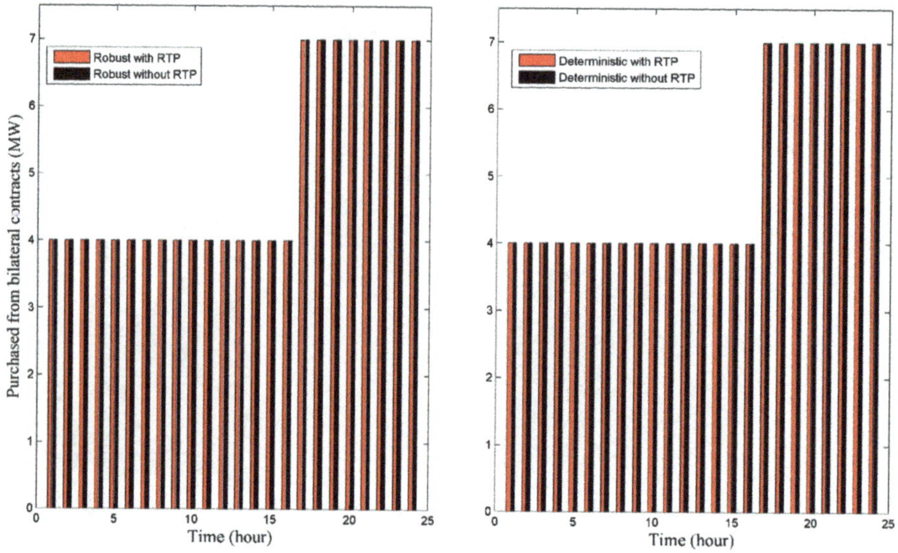

Fig. 9.8 Procured power from bilateral contracts for deterministic and robust cases under implementing RTP

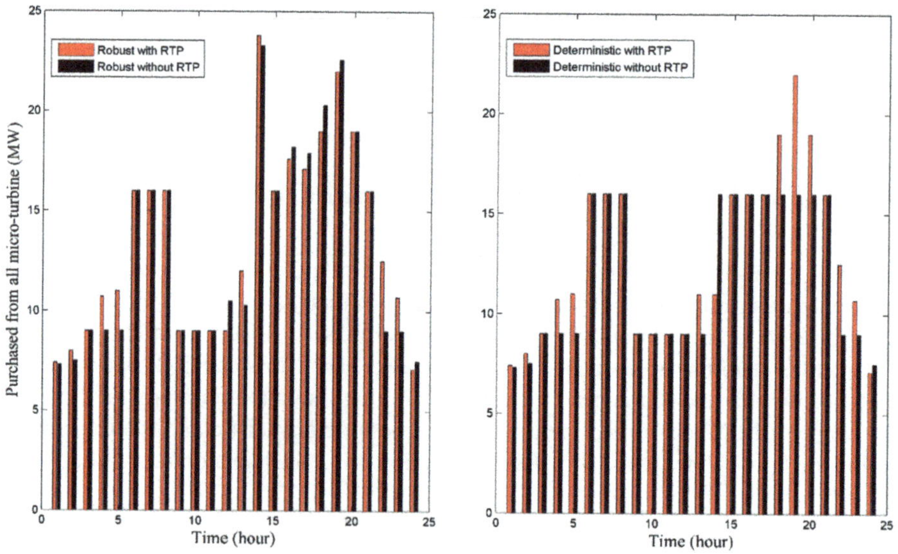

Fig. 9.9 Procured power from self-generation units for deterministic and robust cases under implementing RTP

consumer, which has made the large consumer robust against power price volatility in the pool market. In the robust optimization, under RTP-DRP, the maximum and minimum generated powers by self-generating units are equal to 23.8 and 7.1 MW at hours 14 and 24, respectively. In without RTP case, the maximum and minimum procured powers are obtained as 23.4 and 7.3 MW at hours 14 and 1, respectively. The total generated power by the self-generating units is equal to 316.4 and 322.0 MWh for without and with RTP cases, respectively.

In the deterministic optimization, the maximum and minimum generated powers by under RTP-DRP are equal to 22 and 7.1 MW at hours 19 and 24, respectively. The maximum procured power is obtained as 16 MW at hours 6–8 and 16–22. In addition, the minimum procured power is recorded as 7.3 at hour 1. The total generated power by the self-generating units is equal to 288.30 and 306.40 MWh for without and with RTP cases, respectively.

9.5.5 Charging/Discharging of the Battery Storage System

The charging or discharging state of the battery is illustrated in Fig. 9.10. The maximum charged power of the battery is equal to 0.5 MW, while 0.6 MW is the maximum discharged power. In robust optimization under RTP-DRP, the battery storage is charged at hours 1–3 and 9–12, while it is discharged at hours 7, 14, 18, 19, and 20.

In the deterministic case, the battery is discharged at hours 7 and 18–20. In addition the storage is charged at hours 1–3 and 9–12 in with RTP case. In without RTP case, hours 1–5 are the charged period, and hours 14–15 and 21 are the discharged periods.

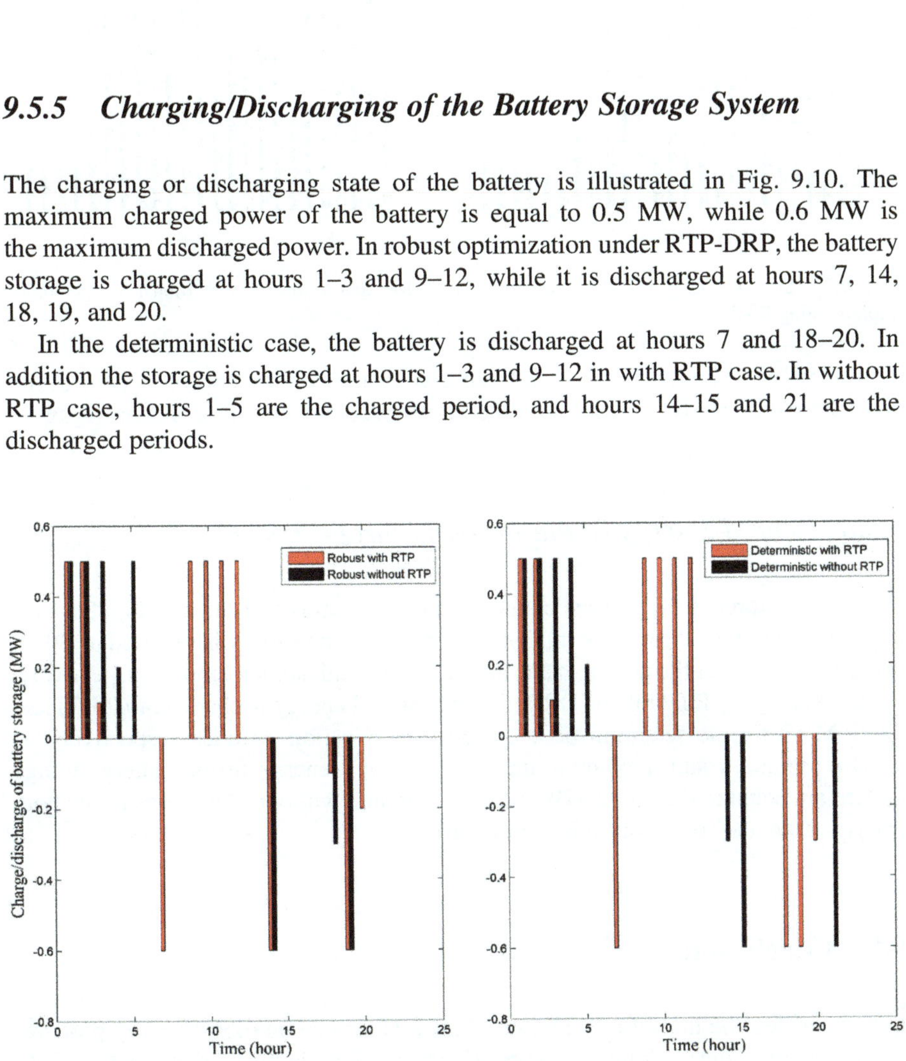

Fig. 9.10 Charging/discharging state of the storage system for deterministic and robust cases under implementing RTP

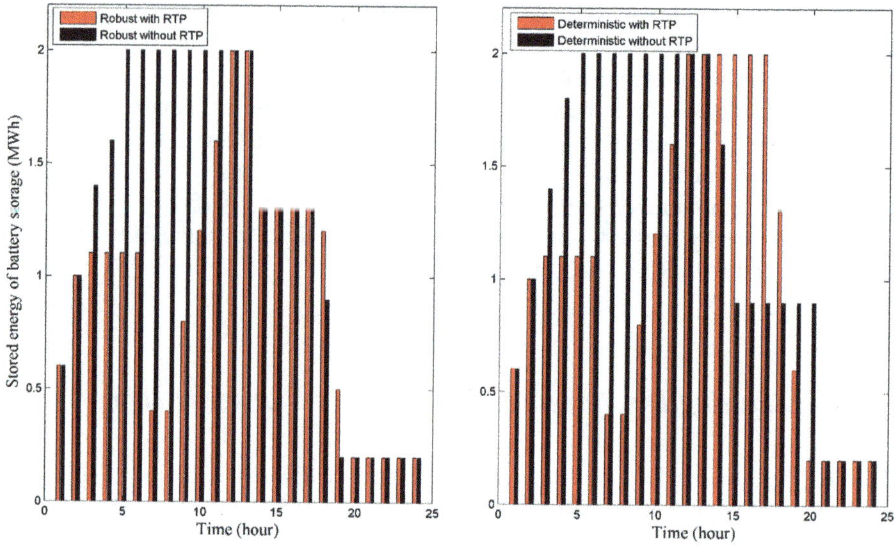

Fig. 9.11 Stored energy of the storage system for deterministic and robust cases under implementing RTP

As said in Chap. 8, due to small capacity of the battery storage system in the considered case study, its impact on the power procurement process is negligible.

9.5.6 Stored Energy in the Battery Storage System

Finally, the stored energy in the battery storage system is presented in Fig. 9.11 for 24 h. In without RTP case, the total stored energy in the battery is reported as 30.60 and 29.90 MWh for the deterministic and robust optimization methods, respectively.

After applying RTP rate of DRP, the total stored energy is changed to 25.30 and 22.30 MWh for the deterministic and robust optimization methods, respectively.

The minimum and maximum amounts of stored energy in the battery storage system are equal to 0.2 and 2 MW for both with and without TOU cases in the both deterministic and robust optimization methods.

9.6 Conclusion

In this chapter, the impact of RTP rate of DRP is studied on robust energy procurement of a large consumer under power price uncertainty in the market in which 30% increase in power price in the pool market is considered and solved using the robust optimization approach. The obtained results are presented for different values of the robustness parameter. Optioned results are compared with deterministic results and robust optimization under time-of-use demand response program which is detailed in Chap. 8.

In the worst-case analysis of robust optimization, total procurement cost is obtained as $41,235.755 without implementing RTP which is reduced to $40,526.599 after applying RTP-DRP indicating 1.72% reduction in total robust power procurement cost. In the deterministic optimization, the total power price is recorded as $40,585.453 and $39,692.237 for without and with RTP cases, respectively, which show 2.2% reduction. Based on the obtained results, it can be inferred that the large consumer is highly robust against the power price uncertainty in the pool market.

The power price uncertainty in the pool market has reduced procured power from the pool market in which by incrusting power price in the market, procured power from the market is reduced. In the deterministic case results, the total cost of procured power from the pool market is obtained as $15,430.565 and $12,932.037 for without and with RTP, respectively, indicating 16.19% reduction. In the worst-case results of robust simulation, the cost of procured energy is obtained as $13,239.416 and $12,153.746 for without and with RTP. It is seen that by implementing RTP, procured power from the pool market is reduced about 8.2%.

The purchased power from the bilateral contracts is obtained as $9322.080 for robust and deterministic optimizations. The minimum purchased power from the bilateral contracts is equal to 4 MW for both optimization methods in with and without RTP cases. In addition, the maximum procured power is equal to 7 MW recorded for both deterministic and robust optimization methods.

The self-generating units supply a significant part of the required load demand of the large consumer. In determinist case, total operating cost of self-generating units is recorded as $15,832.808 and $17,438.120 for without and with RTP, respectively. It is obvious that in with RTP case, the operating cost is higher than without RTP case because more power is required for both cases. In the worst case, results are obtained as $18,674.259 and $19,050.773 for without and with RTP cases, respectively. Finally, the procured power from each of the alternative sources is reported for robust and deterministic optimization methods in without and with RTP cases.

References

1. H. Wang, H. Fang, X. Yu, S. Liang, How real time pricing modifies Chinese households' electricity consumption. J. Clean. Prod. **178**, 776–790 (2018)
2. N. Ghadimi, A. Afkousi-Paqaleh, A. Emamhosseini, A PSO-based fuzzy long-term multi-objective optimization approach for placement and parameter setting of UPFC. Arab. J. Sci. Eng. **39**(4), 2953–2963 (2014)
3. N. Ghadimi, MDE with considered different load scenarios for solving optimal location and sizing of shunt capacitors. Natl. Acad. Sci. Lett. **37**(5), 447–450 (2014)
4. F. Kühnlenz, P.H.J. Nardelli, S. Karhinen, R. Svento, Implementing flexible demand: real-time price vs. market integration. Energy **149**, 550–565 (2018)

5. S. Favuzza et al., Real-time pricing for aggregates energy resources in the Italian energy market. Energy **87**, 251–258 (2015)
6. N. Nezamoddini, Y. Wang, Real-time electricity pricing for industrial customers: survey and case studies in the United States. Appl. Energy **195**, 1023–1037 (2017)
7. Y. Dai, Y. Gao, H. Gao, H. Zhu, Real-time pricing scheme based on Stackelberg game in smart grid with multiple power retailers. Neurocomputing **260**, 149–156 (2017)
8. H. Anand, R. Ramasubbu, A real time pricing strategy for remote micro-grid with economic emission dispatch and stochastic renewable energy sources. Renew. Energy **127**, 779–789 (2018)
9. T. Niknam, A. Kavousifard, S. Tabatabaei, J. Aghaei, Optimal operation management of fuel cell/wind/photovoltaic power sources connected to distribution networks. J. Power Sources **196**(20), 8881–8896 (2011)
10. H.A. Aalami, S. Nojavan, Energy storage system and demand response program effects on stochastic energy procurement of large consumers considering renewable generation. IET Gener. Transm. Distrib. **10**(1), 107–114 (2016)
11. S. Nojavan, H. allah Aalami, Stochastic energy procurement of large electricity consumer considering photovoltaic, wind-turbine, micro-turbines, energy storage system in the presence of demand response program. Energy Convers. Manag. **103**, 1008–1018 (2015)

Chapter 10
Hybrid Heating and Power Energy Procurement

Hossein Khodaei

Abstract For a large industrial consumer, heating energy may be required besides the electricity energy. In this case, multi-carrier energy systems or energy hubs can be considered as a flexible way for energy management of the large consumer. In this chapter, optimal operation of multi-carrier energy systems is studied for a large consumer using different heat and power generating sources such as combined heat and power unit and wind farm, thermal and electrical storages, and thermal and electricity energy markets. Also, to reduce energy procurement cost, demand response programs are considered. The uncertainty of heat and electrical prices, wind speed, and load demands is modeled using stochastic programming method considering a set of discrete scenarios. The normal distribution is used to generate scenarios for heat and electrical load demands and prices, while the Weibull distribution is used to provide scenarios of wind speed in order to model power output of the wind turbine.

10.1 Introduction

In an industrial manufacturing process, the heating energy may be required by the large consumer besides the electricity energy, which by considering it can be a challenge for consumers. To cope with this problem, the concept of energy hub has recently emerged [1, 2] which investigates the whole of energy systems using new technologies such as combined heat and power (CHP). The optimal scheduling of the energy hub systems during different periods is the main challenge in the management of energy hub operation [3, 4].

In this chapter, the optimal scheduling and operation of an energy hub including thermal and electricity energy markets, thermal storages, and wind farm under demand response program and various uncertainties in the system are investigated for a large consumer. The demand response program is implemented for thermal and

H. Khodaei (✉)
Department of Engineering, Ardabil Branch, Islamic Azad University, Ardabil, Iran

S. Nojavan et al. (eds.), *Robust Energy Procurement of Large Electricity Consumers*,
https://doi.org/10.1007/978-3-030-03229-6_10

electrical demand to reduce the operating cost of energy hub through shifting the load into low-price periods [5].

To get realistic model of energy hub for the large consumer, different uncertainties such as electrical and heat demand, natural gas demand, market price of the electrical and thermal energies, and power output of wind farm are taken into account [6]. The problem is formulated as mixed-integer linear programming, and the global optimum operating point obtained solving the problem under CPLEX solver [7] in the GAMS optimization software [8].

10.2 Model of Energy Hub

In order to meet the thermal and electrical demands of a large consumer, natural gas, electricity, and district heating are assumed as input energy carriers, and the heat, electrical, and natural gas demands are considered as the output side in the developed energy hub model. The required energy is generated using multiple devices as CHP, thermal storage, boiler, electrical storage, electrical transformer, and DRP units. The operator of energy hub should determine the optimal operation of different units based on the obtained data from the thermal and electricity markets. Note that all players in the market are assumed price takers.

10.3 Uncertainty Modeling

Related uncertainties of the energy hub operation are modeled using the stochastic programming which is detailed in Chap. 3. The stochastic programming is widely used to model uncertainties of renewable energy [9] and power market price [10]. The problem is formulated considering generated scenarios for aforementioned uncertainties. Note that scenarios of wind farm output are generated using the Weibull distribution, while other uncertainties are modeled based on the normal distribution.

10.4 Formulation of the Problem

The stochastic formulation of daily scheduling problem of energy hub for a large consumer is formulated which is developed in the following subsections.

10.4.1 The Objective Function

The objective function of the problem is solved to minimize the operating cost of the considered energy hub. To do so, Eq. (10.1) presents the objective function including imported natural gas, electricity, district heating, and operation cost of storage systems under the demand response programs:

$$\mathrm{OF}=\sum_{t=1}^{24}\sum_{s}\omega_s\{\cos t_E+\cos t_G+\cos t_H+\cos t_{\mathrm{ES}}$$
$$+\cos t_{\mathrm{HS}}+\cos t_{\mathrm{EDRP}}+\cos t_{\mathrm{TDRP}}\} \quad (10.1)$$

where:

$$\cos t_E=\left[\pi_{\mathrm{net}}^E(t,s)P_{\mathrm{net}}^E(t,s)\right]+\left[\pi_{\mathrm{wind}}^E P_{\mathrm{wind}}^{\mathrm{E}}(t,s)\right] \quad (10.2)$$

$$\cos t_G=\left[\pi_{\mathrm{net}}^G P_{\mathrm{net}}^G(t,s)\right] \quad (10.3)$$

$$\cos t_H=\left[\pi_{\mathrm{net}}^H(t,s)P_{\mathrm{net}}^H(t,s)\right] \quad (10.4)$$

$$\cos t_{\mathrm{ES}}=\left[\pi_{\mathrm{op}}^E\left(P_{\mathrm{ch}}^E(t,s)+P_{\mathrm{dis}}^E(\mathrm{t},s)\right)\right]+\left[\pi_{\mathrm{net}}^E(t,s)\left(P_{\mathrm{ch}}^E(t,s)-P_{\mathrm{dis}}^E(\mathrm{t},s)\right)\right] \quad (10.5)$$

$$\cos t_{\mathrm{HS}}=\left[\pi_{\mathrm{op}}^H\left(P_{\mathrm{ch}}^H(t,s)+P_{\mathrm{dis}}^H(t,s)\right)\right]+\left[\pi_{\mathrm{net}}^H(t,s)\left(P_{\mathrm{ch}}^H(t,s)-P_{\mathrm{dis}}^H(t,s)\right)\right] \quad (10.6)$$

$$\cos t_{\mathrm{EDRP}}=\left[\pi_{\mathrm{DR}}^E\left(P_{\mathrm{down}}^E(t,s)+P_{\mathrm{up}}^E(t,s)\right)\right] \quad (10.7)$$

$$\cos t_{\mathrm{TDRP}}=\left[\pi_{\mathrm{DR}}^H\left(P_{\mathrm{down}}^H(t,s)+P_{\mathrm{up}}^H(t,s)\right)\right] \quad (10.8)$$

where:

$\pi_{\mathrm{net}}^E(t,s)/\pi_{\mathrm{net}}^H(t,s)$ is the price of electricity/heat in scenario s and time t.
π_{wind}^E is the operating cost of wind turbine.
π_{net}^G is the price of natural gas.
$\pi_{\mathrm{op}}^E/\pi_{\mathrm{op}}^H$ is the operating cost of electrical/heat storage.
$\pi_{\mathrm{DR}}^E/\pi_{\mathrm{DR}}^H$ is the cost of electrical/thermal demand response program.

In Eq. (10.1), ω_s is the probability of scenario s. Equation (10.2) presents the cost of purchased electricity including purchased power from the power market and wind farm. The cost of purchased gas and district heating cost are provided by Eqs. (10.3) and (10.4), respectively. The operating cost of electrical and heat storages is modeled by Eqs. (10.5) and (10.6), respectively. Finally, the cost of implementing demand response programs is presented by Eqs. (10.7) and (10.8), respectively.

In the abovementioned formulations

$P^E_{\text{net}}(t,s)/P^H_{\text{net}}(t,s)/P^G_{\text{net}}(t,s)$ is the amount of purchased electrical/heat/gas from the market at scenario s and time t

$P^E_{\text{ch}}(t,s)/P^E_{\text{dis}}(t,s)$ is the amount of charged/discharged electricity power of storage in scenario s and time t

$P^H_{\text{ch}}(t,s)/P^H_{\text{dis}}(t,s)$ is the amount of charged/discharged thermal energy of the storage in scenario s and time t

$P^E_{\text{down}}(t,s)/P^E_{\text{up}}(t,s)$ is the amount of shifted down/up electrical power by demand response program

$P^H_{\text{down}}(t,s)/P^H_{\text{up}}(t,s)$ is the amount of shifted down/up thermal energy by demand response program

$P^E_{\text{wind}}(t,s)$ is the procured power from wind turbine in scenario s and time t

10.4.2 Constraints

In the following subsections, required constraints for modeling the energy hub such as energy balance and operational constraint are presented.

10.4.2.1 Energy Balance

The required load demand by the large consumer should be supplied by the energy hub considering different types of energy carriers. Equation (10.9) presents the electricity energy balance:

$$\begin{aligned} P^E_{\text{demand}}(t,s) = {} & \left[\eta^{EE}_{\text{Trans}} P^E_{\text{net}}(t,s)\right] + \left[\eta^{\text{GE}}_{\text{CHP}} P^G_{\text{netCHP}}(t,s)\right] + \left[\eta^{\text{EE}}_{\text{Conv}} P^{\text{E}}_{\text{wind}}(t,s)\right] \\ & + \left[P^E_{\text{dis}}(t,s) - P^E_{\text{ch}}(t,s)\right] + \left[P^E_{\text{down}}(t,s) - P^E_{\text{up}}(t,s)\right] \end{aligned} \tag{10.9}$$

where:

$P^E_{\text{demand}}(t,s)$ is the electricity demand of energy hub in scenario s and time t.
P^G_{netCHP} is the purchased natural gas for CHP unit in scenario s and time t.
$\eta^{\text{EE}}_{\text{Trans}}$ is the efficiency of the electrical transformer.
$\eta^{\text{GE}}_{\text{CHP}}$ is the electricity efficiency of CHP unit.
$\eta^{\text{EE}}_{\text{Conv}}$ is the efficiency of intermediate converter of the wind turbine.

The balance between purchased natural gas from the market and consumed amount of gas is provided in Eq. (10.10):

$$P^G_{\text{demand}}(t,s) = P^G_{\text{net}}(t,s) - P^G_{\text{netCHP}}(t,s) - P^G_{\text{netB}}(t,s) \tag{10.10}$$

In Eq. (10.10), $P^{E}_{\text{demand}}(t,s)$ is the natural gas demand of energy hub in scenario s and time t, and $P^{G}_{\text{netB}}(t,s)$ is natural gas power input of boiler in scenario s and time t.

Balancing between the heat energy carriers is provided by Eq. (10.11). In this chapter, heat demand is modeled as an energy demand, which is provided by thermal energy generators of the hub:

$$P^{H}_{\text{demand}}(t,s) = \left[\eta^{\text{GH}}_{\text{CHP}} P^{G}_{\text{netCHP}}(t,s)\right] + \left[\eta^{\text{GH}}_{\text{Boil}} P^{G}_{\text{netB}}(t,s)\right] + \left[P^{H}_{\text{dis}}(t,s) - P^{H}_{\text{ch}}(t,s)\right] \tag{10.11}$$

where $P^{H}_{\text{demand}}(t,s)$ is the heat demand of energy hub and $\eta^{\text{GH}}_{\text{CHP}}/\eta^{\text{GH}}_{\text{Boil}}$ is the gas to heat efficiency of the CHP/boiler.

10.4.2.2 Operational Requirements

In this section, the operational constraints of different devices in the hub are provided. Equations (10.12)–(10.14) satisfy the connection point constraints, and constraints (10.14)–(10.17) avoid the devices form operating in unallowable ranges:

$$0 \le P^{E}_{\text{net}}(t,s) \le P^{E}_{\text{netmax}} \tag{10.12}$$

$$0 \le P^{H}_{\text{net}}(t,s) \le P^{H}_{\text{netmax}} \tag{10.13}$$

$$0 \le P^{G}_{\text{net}}(t,s) \le P^{G}_{\text{netmax}} \tag{10.14}$$

$$0 \le P^{E}_{\text{net}}(t,s) \le P^{\text{input}}_{\text{Trans}} \tag{10.15}$$

$$0 \le P^{G}_{\text{netCHP}}(t,s) \le P^{\text{input}}_{\text{CHP}} \tag{10.16}$$

$$0 \le P^{G}_{\text{netB}}(t,s) \le P^{\text{input}}_{\text{boiler}} \tag{10.17}$$

where

$P^{E}_{\text{netmax}}/P^{H}_{\text{netmax}}/P^{G}_{\text{netmax}}$ is the maximum capacity of purchased power/heat/district heat from the market. Equation (10.12) presents the electrical limitation. Constraint (10.13) which is developed based on the design of district heating network connection point indicates the purchased energy from district heating network. Equation (10.14) provides the upper bound of natural gas carrier. Finally, by using constraints (10.15)–(10.17), the operation of devices is satisfied in their permitted range.

10.4.2.3 Electrical Storage

Equations (10.18)–(10.23) present the modeling of the electrical storage system in which Eq. (10.18) provides the state of charge of electrical storage, Eq. (10.19) models the loss of electrical storage, Eq. (10.20) keeps the storage energy level in the allowed range, and Eqs. (10.21) and (10.22) provide storage charging and discharging limitation, respectively. Finally, simultaneously charging and discharging of the storage system is prevented by Eq. (10.23):

$$P_s^E(t,s) = P_s^E(t-1,s) + P_{\text{ch}}^E(t,s) - P_{\text{dis}}^E(t,s) - P_{\text{loss}}^E(t,s) \tag{10.18}$$

$$P_{\text{loss}}^E(t,s) = \alpha_{\text{loss}}^E P_s^E(t,s) \tag{10.19}$$

$$\alpha_{\text{Min}}^E P_{\text{capa}}^E \le P_s^E(t,s) \le \alpha_{\text{Max}}^E P_{\text{capa}}^E \tag{10.20}$$

$$\beta_{\text{Min}}^E P_{\text{capa}}^E \left(1/\eta_{\text{ES}}^{\text{ch}}\right) I_{\text{ch}}^E(t,s) \le P_{\text{ch}}^E(t,s) \le \beta_{\text{Max}}^E P_{\text{capa}}^E \left(1/\eta_{\text{ES}}^{\text{ch}}\right) I_{\text{ch}}^E(t,s) \tag{10.21}$$

$$\beta_{\text{Min}}^E P_{\text{capa}}^E \eta_{\text{ES}}^{\text{dis}} I_{\text{ch}}^E(t,s) \le P_{\text{dis}}^E(t,s) \le \beta_{\text{Max}}^E P_{\text{capa}}^E \eta_{\text{ES}}^{\text{dis}} I_{\text{ch}}^E(t,s) \tag{10.22}$$

$$0 \le I_{\text{dis}}^E(t,s) + I_{\text{ch}}^E(t,s) \le 1 \tag{10.23}$$

where:

α_{loss}^E is the loss ratio of the electrical storage.
$\alpha_{\text{Min}}^E/\alpha_{\text{Max}}^E$ is the minimum/maximum ratios of power level of electrical storage.
P_{capa}^E is the capacity of the electrical storage.
$\beta_{\text{Min}}^E/\beta_{\text{Max}}^E$ is the minimum/maximum allowable rate of charging the electrical storage.
$\eta_{\text{ES}}^{\text{ch}}/\eta_{\text{ES}}^{\text{dis}}$ is the charging/discharging efficiency of the electrical storage.

10.4.2.4 Modeling of Thermal Storage

The operational constraints of the thermal storage system are modeled using Eqs. (10.24)–(10.29). Equation (10.24) models the state of charge of thermal storage, Eq. (10.25) models the thermal loss of thermal storage system, the energy level is kept in allowed operation range by Eq. (10.26), and Eqs. (10.27) and (10.28) provide the rate of charging and discharging of storage system. Finally, constraint (10.29) prevents simultaneous charging and discharging of the thermal storage:

$$P_s^H(t,s) = P_s^H(t-1,s) + P_{\text{ch}}^H(t,s) - P_{\text{dis}}^H(t,s) - P_{\text{loss}}^H(t,s) \tag{10.24}$$

$$P_{\text{loss}}^H(t,s) = \alpha_{\text{loss}}^H P_s^H(t,s) \tag{10.25}$$

$$\alpha_{\text{Min}}^H P_{\text{capa}}^H \le P_s^H(t,s) \le \alpha_{\text{Max}}^H P_{\text{capa}}^H \tag{10.26}$$

$$\beta_{\text{Min}}^H P_{\text{capa}}^H \left(1/\eta_{\text{HS}}^{\text{ch}}\right) I_{\text{ch}}^E(t,s) \le P_{\text{ch}}^H(t,s) \le \beta_{\text{Max}}^H P_{\text{capa}}^H \left(1/\eta_{\text{ES}}^{\text{ch}}\right) I_{\text{ch}}^H(t,s) \tag{10.27}$$

$$\beta_{\text{Min}}^H P_{\text{capa}}^H \eta_{\text{HS}}^{\text{dis}} I_{\text{ch}}^H(t,s) \le P_{\text{dis}}^H(t,s) \le \beta_{\text{Max}}^H P_{\text{capa}}^H \eta_{\text{HS}}^{\text{dis}} I_{\text{dis}}^H(t,s) \tag{10.28}$$

$$0 \le I_{\text{dis}}^{H}(t,s) + I_{\text{ch}}^{H}(t,s) \le 1 \tag{10.29}$$

where:

α_{loss}^{H} is the loss ratio of the thermal storage.
$\alpha_{\text{Min}}^{H}/\alpha_{\text{Max}}^{H}$ is the minimum/maximum ratios of power level of electrical storage.
P_{capa}^{H} is the capacity of the thermal storage.
$\beta_{\text{Min}}^{H}/\beta_{\text{Max}}^{H}$ is the minimum/maximum allowable rate of charging the thermal storage.
$\eta_{\text{HS}}^{\text{ch}}/\eta_{\text{HS}}^{\text{dis}}$ is the charging/discharging efficiency of the thermal storage.

10.4.2.5 Constraints of the Wind Farm

The power output of wind turbine depends on the wind speed. The output power of the wind turbine is modeled using Eq. (10.30) [11]:

$$P_{\text{wind}}^{E}(v(t,s)) = \begin{cases} 0 & \text{if } v(t) \le v_{\text{in}}^{c} \text{ or } v(t) \ge v_{\text{out}}^{c} \\ \dfrac{v(t) - v_{\text{in}}^{c}}{v_{\text{rated}}^{c} - v_{\text{in}}^{c}} & \text{if } v_{\text{in}}^{c} \le v(t) \le v_{\text{rated}}^{c} \\ P_{r}^{w} & \text{if } v_{\text{rated}}^{c} \le v(t) \le v_{\text{out}}^{c} \end{cases} \tag{10.30}$$

where:

P_{r}^{w} is the rated power output of the wind turbine.
$v(t)$ is the wind speed in scenario s and time t.
$v_{\text{in}}^{c}/v_{\text{out}}^{c}$ is the cut-in/cutout speed of the wind turbine.
v_{rated}^{c} is the rated speed.

10.4.2.6 The Electrical Demand Response Program Constraints

The electrical demand response program is considered as useful tool for decreasing operation cost of the system. The electrical demand response program formulation is provided in Eqs. (10.31)–(10.34) [12]. Equation (10.31) presents the equality of shifted up and down electrical loads; Eqs. (10.32) and (10.33) present the upper limits of shifted up and down load, respectively. Finally, constraint (10.34) prevents the simultaneous shifting up and down:

$$\sum_{t=1}^{24} P_{\text{down}}^{E}(t,s) = \sum_{t=1}^{24} P_{\text{up}}^{E}(t,s) \tag{10.31}$$

$$0 \le P_{\text{up}}^{E}(t,s) \le \text{LPF}_{\text{up}}^{E} P_{\text{demand}}^{E}(t,s) I_{\text{up}}^{E}(t,s) \tag{10.32}$$

$$0 \le P_{\text{down}}^{E}(t,s) \le \text{LPF}_{\text{down}}^{E} P_{\text{demand}}^{E}(t,s) I_{\text{down}}^{E}(t,s) \tag{10.33}$$

$$0 \le I_{\text{up}}^{E}(t,s) + I_{\text{down}}^{E}(t,s) \le 1 \tag{10.34}$$

where $\text{LPF}_{\text{up}}^{E}/\text{LPF}_{\text{down}}^{E}$ is the maximum ratio of shifted up/down electrical demand.

10.4.2.7 The Thermal Demand Response Program Constraints

The demand response program can be applied on the thermal load as considered for electrical load in the previous section. The formulation of thermal demand response program is provided by Eqs. (10.35)–(10.38). Constraint (10.35) presents the equality of shifted up and down thermal loads, and the upper limits of shifted up and down load are provided by Eqs. (10.36) and (10.37), respectively. Finally, constraint (10.38) prevents the simultaneous shifting up and down:

$$\sum_{t=1}^{24} P_{\text{down}}^{H}(t,s) = \sum_{t=1}^{24} P_{\text{up}}^{H}(t,s) \tag{10.35}$$

$$0 \le P_{\text{up}}^{H}(t,s) \le \text{LPF}_{\text{up}}^{H} P_{\text{demand}}^{H}(t,s) I_{\text{up}}^{H}(t,s) \tag{10.36}$$

$$0 \le P_{\text{down}}^{H}(t,s) \le \text{LPF}_{\text{down}}^{H} P_{\text{demand}}^{H}(t,s) I_{\text{down}}^{H}(t,s) \tag{10.37}$$

$$0 \le I_{\text{down}}^{H}(t,s) + I_{\text{up}}^{H}(t,s) \le 1 \tag{10.38}$$

where $\text{LPF}_{\text{up}}^{H}/\text{LPF}_{\text{down}}^{H}$ is the maximum ratio of shifted up/down electrical demand.

10.5 Simulation Information

The problem is solved considering different uncertainties using the stochastic programming method under electrical and thermal demand response program. In order to investigate the impact of demand response program, the problem two case studies are considered as follows:

Case 1: With implementing electrical and thermal DRP
Case 2: Without electrical and thermal DRP

To model the uncertainty of electricity market, wind farm, and storage systems, ten discrete scenarios are generated in both cases. As said before, the Weibull distribution is used to generate wind speed scenarios, while the normal distribution is used to generate scenarios for modeling the rest of uncertain parameters. Figure 10.1 presents the wind speed for 24 h in scenario 5. Based on the wind speed in scenario 5, the power output of the wind turbine is depicted in Fig. 10.2. The

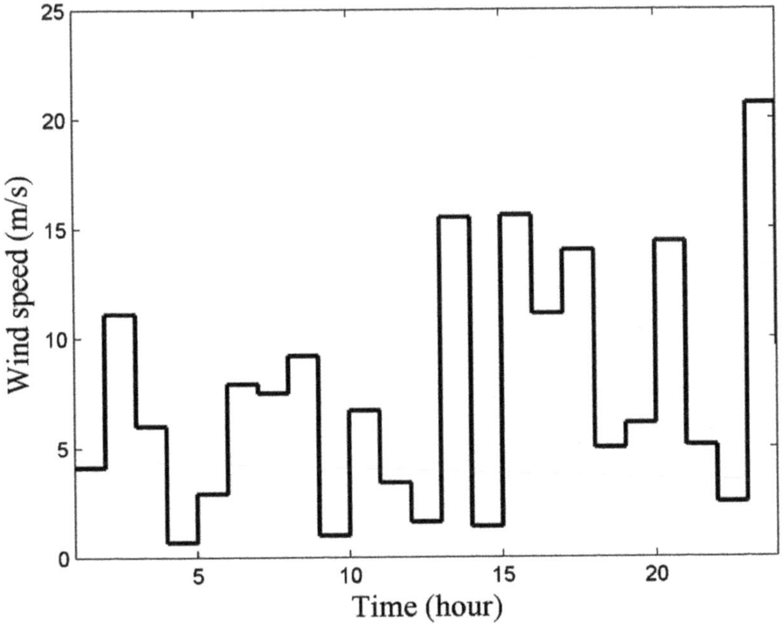

Fig. 10.1 Wind speed in scenario 5

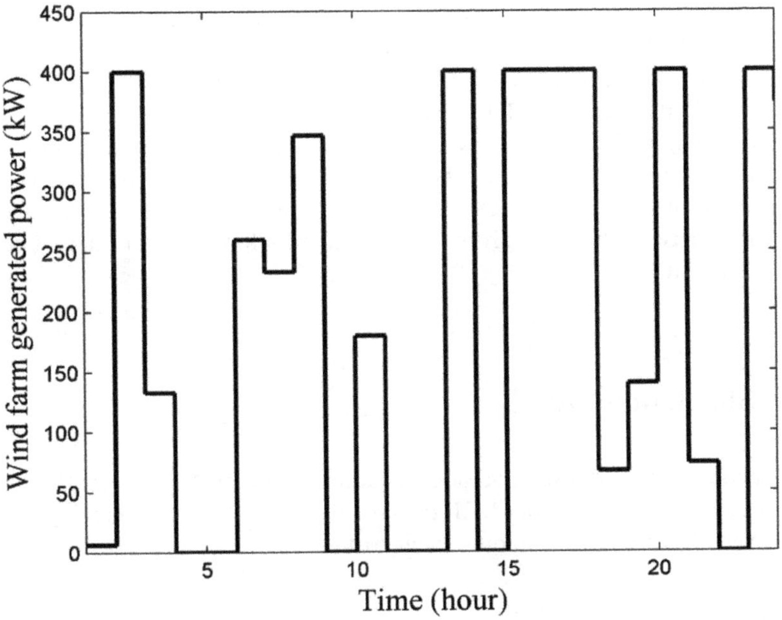

Fig. 10.2 Power output of wind farm in scenario 5

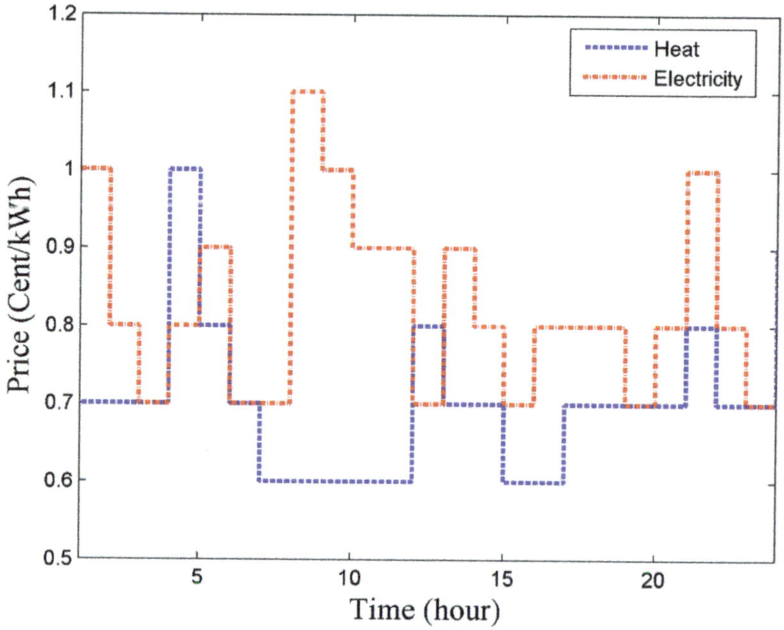

Fig. 10.3 Electricity and thermal market price in scenario 5

maximum power output of the wind farm is obtained as 400 KW at hours 2,13, 15–17, 20, and 23, while the minimum power is 0 at hours 4–6, 9, 11–13, and 2.

The thermal and electricity energy prices are depicted in Fig. 10.3. The maximum price of electricity is equal to 1.1 Cent/Kwh, while the minimum price is 0.7 Cent/kWh. According to Fig. 10.3, the maximum and minimum electricity power prices are 1 and 0.7 Cent/kWh. Note that the results are reported for scenario 5. Thermal, electrical, and natural gas demand of the energy hub is depicted in Fig. 10.4 for deterministic case. Table 10.1 gives all the required information of internal devices of the energy hub [13].

10.6 Simulation Results

In this section, obtained results of the simulation are presented for scenario 5 in two cases as with DRP and without DRP to show the impact of implementing DRP on thermal and electrical energy requirement cost.

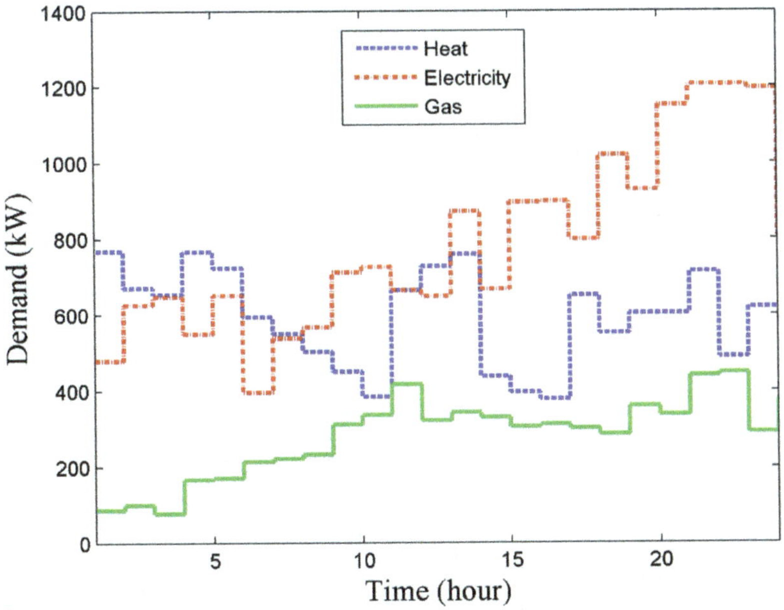

Fig. 10.4 Thermal, electrical, and natural gas demand in scenario 5

10.6.1 Total Procurement Cost

The total procurement cost of the energy hub is equal to \$27,961.0 and \$27,508.0 for with and without DRP indicating 1.62% reduction due to implementing DRP. Note that results are reported for expected procurement cost.

Figures 10.5 and 10.6 show the electrical and thermal load demand for with and without DRP. The electrical DRP has transferred load from high-price periods to low-price periods. For example, the load demand is reduced at hour 8 when the power price is 1.1 Cent/kW, which is the maximum electricity power price. The thermal demand response program has not changed the thermal load profile in scenario 5, meaning that the thermal load profile in this scenario has the optimal distribution in 24 h.

10.6.2 Electrical and Thermal Storages

Electrical and thermal storages can be used to store energy during low-price periods in which using the stored energy during the high-price periods will prevent paying high prices for energy procurement.

Stored electrical energy in the storage is shown in Fig. 10.7 for 24 h and two cases as with and without DRP. The maximum stored energy is equal to 116.7 and

Table 10.1 Required information for simulation

Parameter	Unit	Amount	Parameter	Unit	Amount
π^E_{wind}	Cent/kW h	0	α^H_{Min}	–	0.05
π^G_{net}	Cent/kW h	7.2	α^H_{Max}	–	0.90
π^H_{net}	Cent/kW h	8	β^E_{Min}	–	0.05
π^E_{op}	Cent/kW h	0	β^F_{Max}	–	0.18
π^H_{op}	Cent/kW h	2	β^H_{Min}	–	0.05
π^E_{DR}	Cent/kW h	0	β^H_{Max}	–	0.09
π^H_{DR}	Cent/kW h	0	P^E_{capa}	kW	300
$\eta^{\text{EE}}_{\text{Trans}}$	–	0.90	P^H_{capa}	kW	200
$\eta^{\text{GE}}_{\text{CHP}}$	–	0.4	$\eta^{\text{ch}}_{\text{ES}}$	–	0.09
$\eta^{\text{EE}}_{\text{Conv}}$	–	0.9	$\eta^{\text{dis}}_{\text{ES}}$	–	0.90
$\eta^{\text{GH}}_{\text{CHP}}$	–	0.35	$\eta^{\text{ch}}_{\text{HS}}$	–	0.90
$\eta^{\text{GH}}_{\text{boil}}$	–	0.85	$\eta^{\text{dis}}_{\text{HS}}$	–	0.90
P^E_{netmax}	kW	2000	v^c_{in}	m/s	4
P^G_{netmax}	kW	1800	v^c_{rated}	m/s	10
P^H_{netmax}	kW	2000	v^c_{out}	m/s	22
$P^{\text{input}}_{\text{Trans}}$	kW	2000	P^w_r	kW	400
$P^{\text{input}}_{\text{CHP}}$	kW	800	LPF^E_{up}	–	0.2
$P^{\text{input}}_{\text{boiler}}$	kW	800	LPF^E_{down}	–	0.2
α^E_{loss}	–	0.05	LPF^H_{up}	–	0.2
α^H_{loss}	–	0.05	LPF^H_{down}	–	0.2
α^E_{Min}	–	0.05	Weibull scale	–	8.75
α^E_{Max}	–	0.90	Weibull shape	–	1.75

224.9 kWh at hour 6 and 7 for with and without DRP cases. Also, 15 kW is the minimum amount of stored energy for both cases. The total amount of stored energy in the storage is 1453.2 and 956.9 for with and without DRP cases. The charging and discharging state of the storage at each hour is presented in Fig. 10.8. The maximum charged and discharged energy is equal to 60 and 48.6 kW in both cases, respectively. During charging and discharging process, part of energy can be wasted. Figure 10.9 illustrates the electrical loss of the storage for time horizon of the study in both cases. It is obvious that energy loss flows the pattern of stored energy in the storage. For example, the maximum and minimum amount of energy loss is recorded at same hours that the maximum and minimum amount of electrical energy is stored.

To store the thermal energy, the thermal storage system is considered in the energy hub model. Figure 10.10 presents the stored energy in the storage. It is obvious that stored energy at each hour is the same in both with and without DRP cases. Total amount of stored energy is equal to 359.40 kWh. The maximum and minimum stored energy is equal to 29.6 and 10 kW in both cases, respectively. The

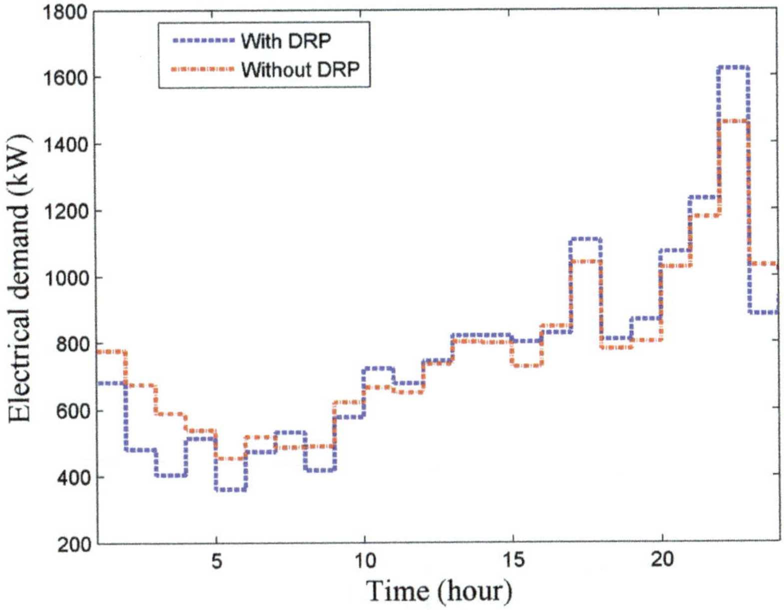

Fig. 10.5 Electrical load demand for with and without DRP in scenario 5

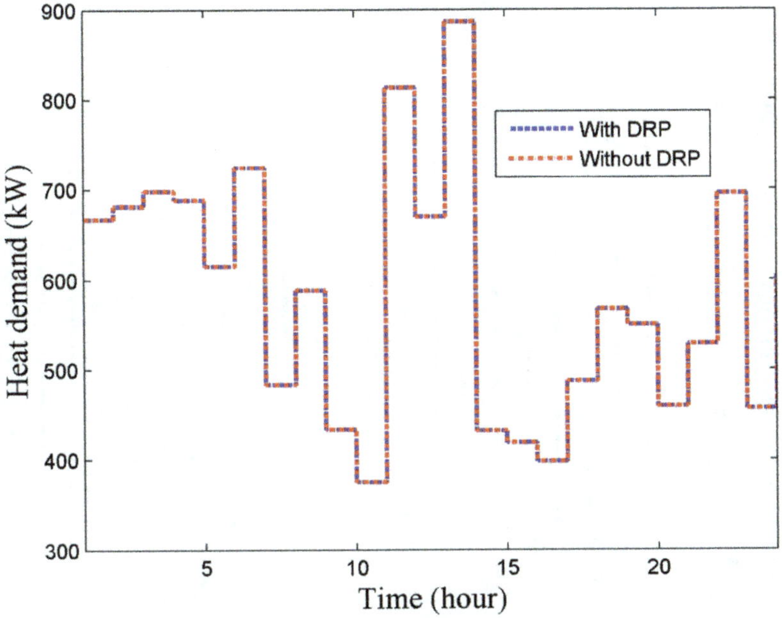

Fig. 10.6 Thermal load demand for with and without DRP in scenario 5

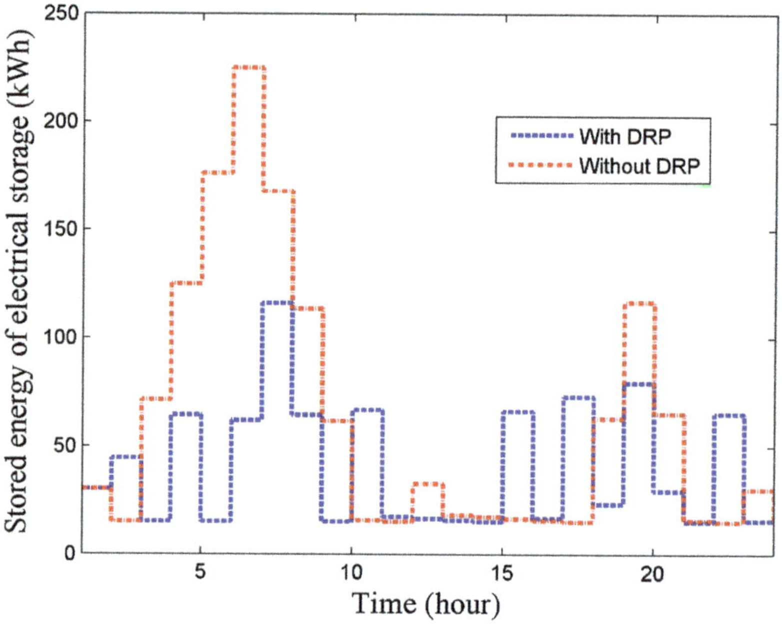

Fig. 10.7 Stored electrical energy in the storage

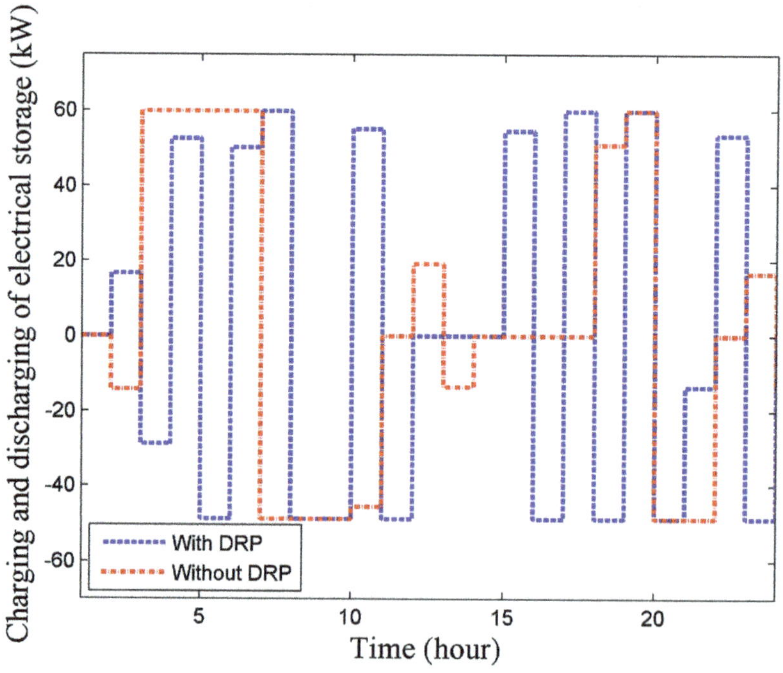

Fig. 10.8 Charging and discharging of the electrical storage

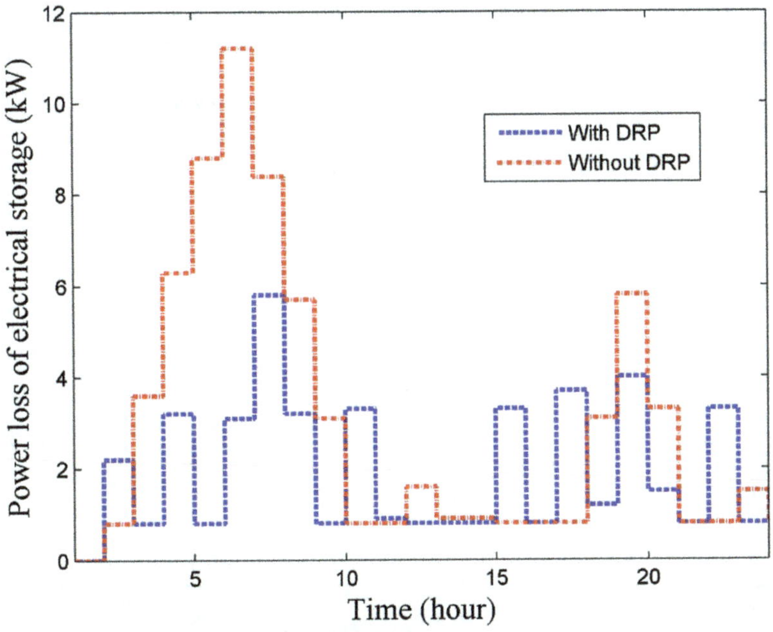

Fig. 10.9 Electricity energy loss of the electrical storage

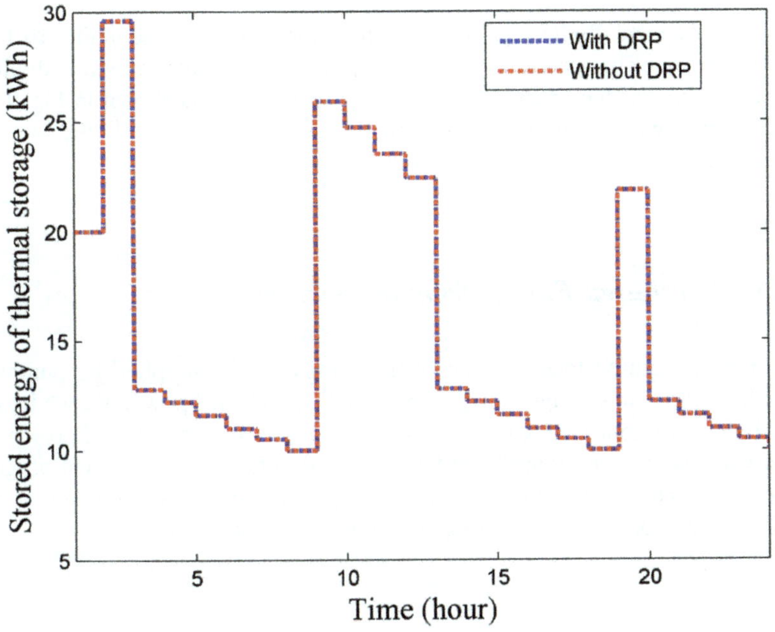

Fig. 10.10 Stored thermal energy in the storage

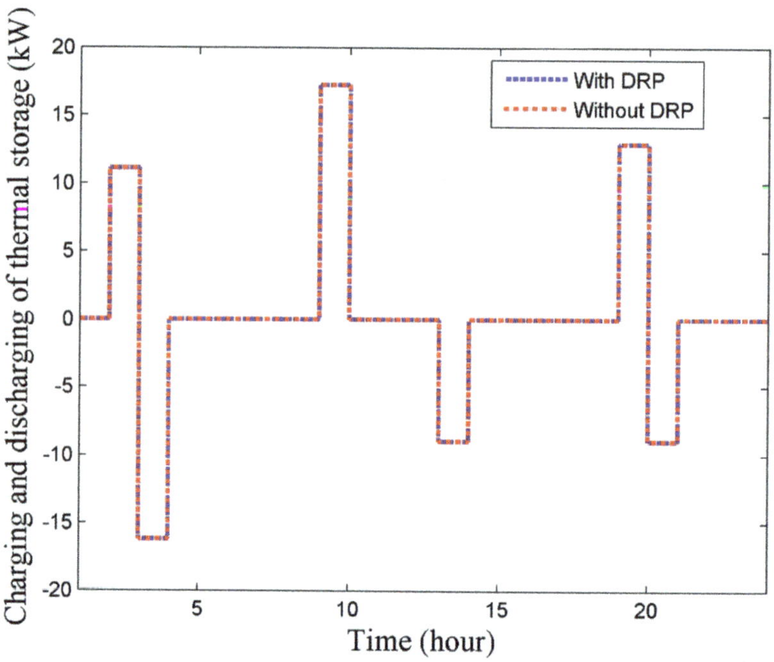

Fig. 10.11 Charging and discharging states of the thermal storage

state of charging and discharging of the thermal storage is provided in Fig. 10.11. As was expected, the charging and discharging power is the same in with and without DRP cases. Finally, the energy loss of the thermal storage is shown in Fig. 10.12. The maximum wasted energy is equal to 1.5 kW at hour 2, while the minimum amount of energy loss is 0 recorded at hour 1.

10.6.3 Purchased Energy from the Market

Some part of required load demand of the energy hub is supplied by purchasing energy from the market. Figure 10.13 presents the purchased power at each hour for both with and without DRP cases. According to Fig. 10.13, purchased power from the market is decreased by applying DRP resulted reduction in total cost of energy procurement. The maximum amount of imported power is recorded at hour 22, which is coincidence with the maximum required load demand.

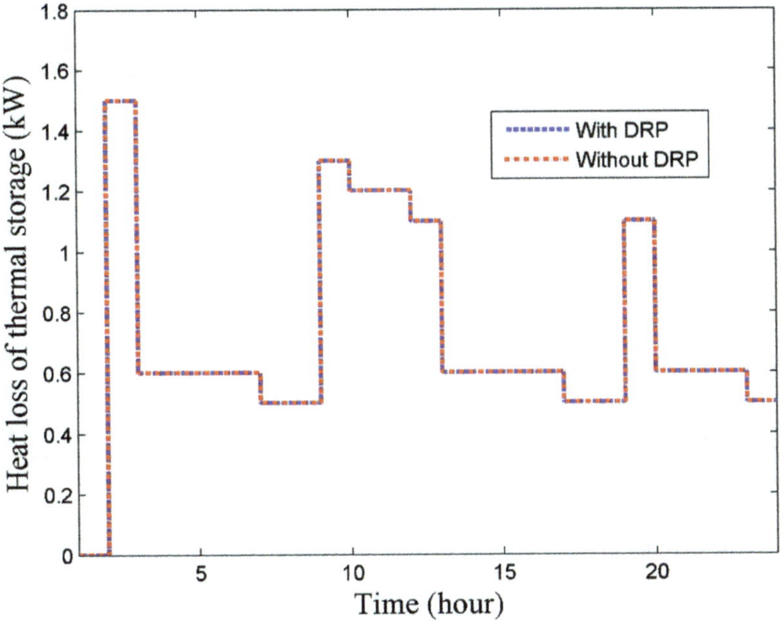

Fig. 10.12 Energy loss of the thermal storage

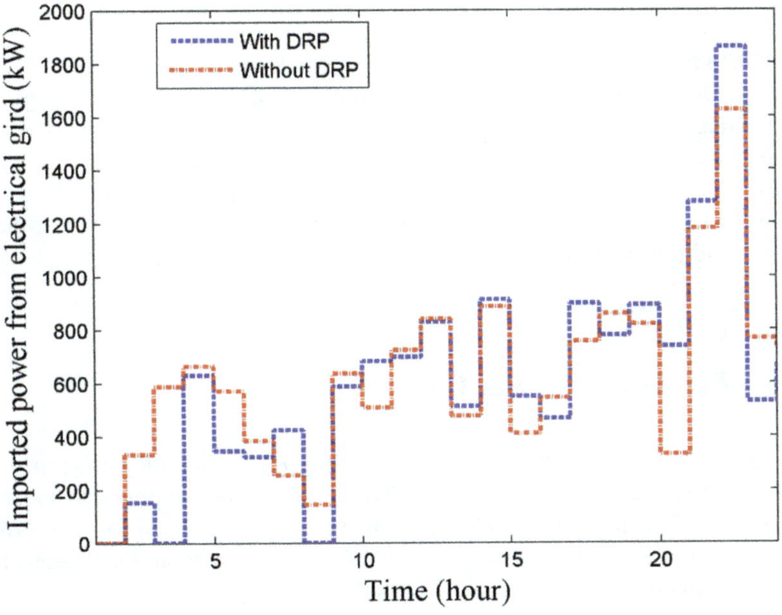

Fig. 10.13 Imported energy from the electrical grid

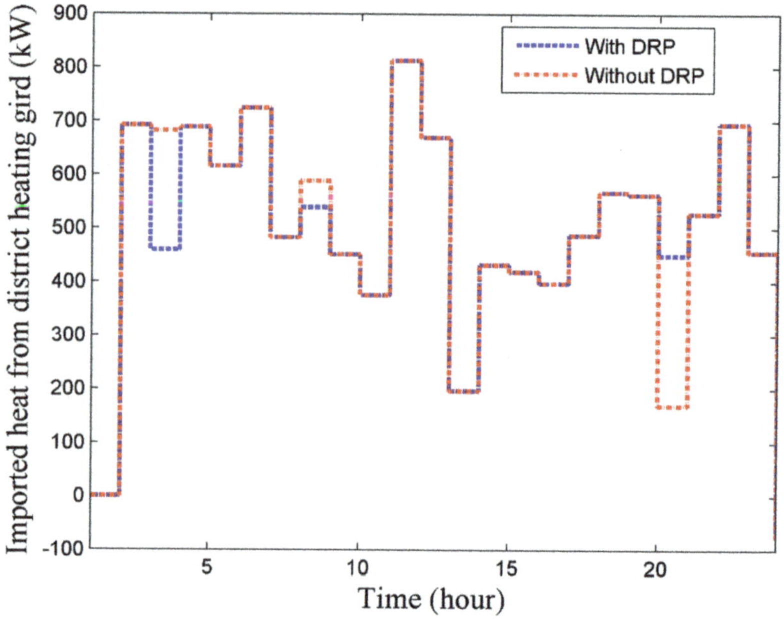

Fig. 10.14 Imported energy from the thermal market

10.6.4 Purchased Thermal Energy from the Market

Figure 10.14 presents the purchased thermal energy from the thermal market to meet the required demand of the hub. It is obvious that purchased energy after implementing the DRP is increased in which total purchased energy is equal to 1.161.6 and 1.1624 MWh for without and with DRP cases, respectively. In without DRP case, the maximum and minimum purchased energy are recorded as 813 and 0 kW at hours 11 and 1, respectively. Note that according to Fig. 10.14, thermal energy is sold to grid at hour 24.

10.6.5 Purchased Natural Gas

Purchased natural gas from the grid is shown in Fig. 10.15. Natural gas is procured to feed heat and power generating units as CHP and boiler and supply gas demand of the energy hub. The maximum and minimum amount of purchased gas is recorded at hours 13 and 1 equal to 1.901 kW and 0 in both with and without DRP cases. In the following subsection, fed gas to the CHP and boiler is provided.

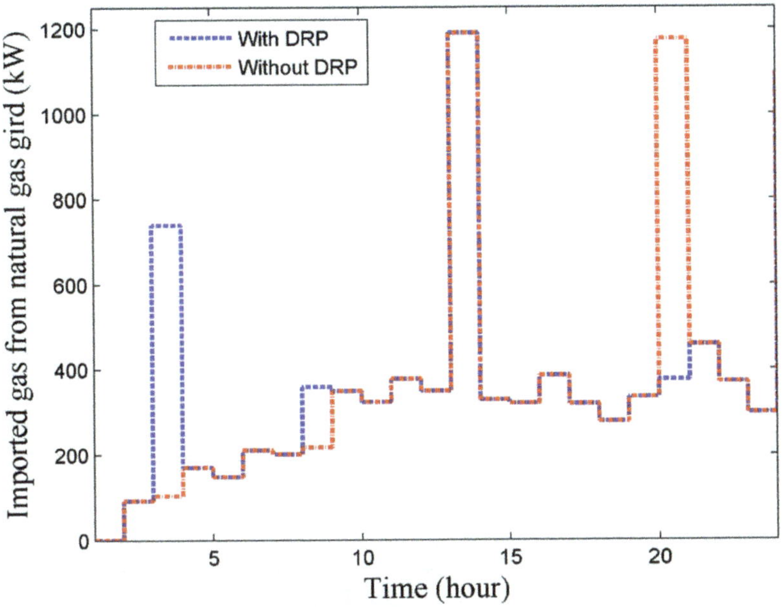

Fig. 10.15 Purchased natural gas from the grid

10.6.5.1 Natural Gas Input of CHP

Figure 10.16 illustrates the gas input of the CHP unit. Obtained results are similar in both with and without DRP cases. According to Fig. 10.16, the natural gas is only inputted at hour 13 to produce heat and power. Total inputted power is equal to 800 kW.

10.6.5.2 Natural Gas Input of CHP

The gas input of the boiler is depicted in Fig. 10.17 for both with and without DRP cases. In without DRP case, total purchased natural gas is equal to 800 kW which is recorded at hour 20, while in other periods the gas input is equal to 0. The maximum purchased natural gas in with DRP case is obtained as 636.3 kW at hour 4. The total gas input of the boiler is equal to 776.1 kWh. It should be reminded that the rest of purchased natural gas from the grid is used to supply the gas demand of the energy hub.

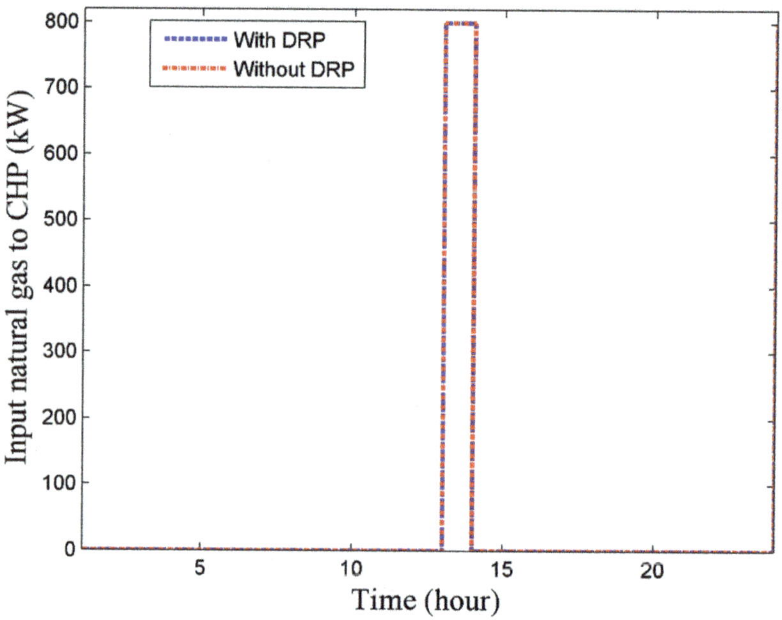

Fig. 10.16 Natural gas input of CHP

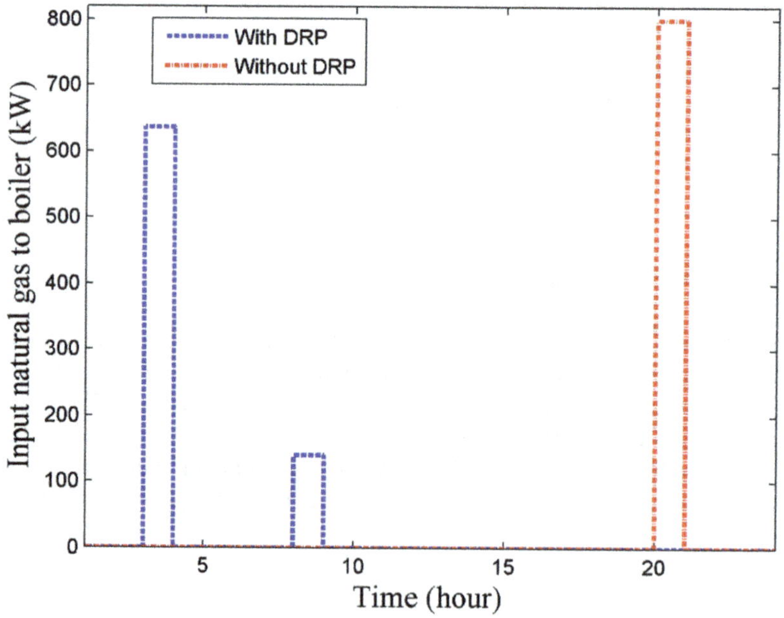

Fig. 10.17 Natural gas input of the boiler

10.7 Conclusion

In this chapter, an energy hub model is developed to meet the heat and power demand of a large consumer considering different energy carriers. In addition, thermal and electricity energy markets, thermal storages, combined heat and power unit, boiler, and wind farm are considered as the main energy procurement sources. The heat and power demand response programs are implemented to reduce energy procurement cost by shifting load from high-price periods to low-price periods. The uncertainties of electrical and heat demand, natural gas demand, market price of the electrical and thermal energies, and power output of wind farm are considered in the optimal scheduling of the energy hub using the stochastic programming method.

To do so, ten discrete scenarios are generated using the normal and Weibull distributions. To show the impact of heat and electrical demand response program, the problem is solved in two cases as with and without DRP.

The problem is formulated as mixed-integer linear programming, and the global optimum operating point obtained solving the problem under CPLEX solver in the GAMS optimization software.

Based on obtained results, the total procurement cost of the energy hub is equal to \$27,961.0 and \$27,508.0 for with and without DRP indicating 1.62% reduction due to implementing DRP. Finally, the performance of generating devices is modeled and investigated during 24 h.

References

1. P. Favre-Perrod, A vision of future energy networks, in *Proceedings of the Inaugural IEEE PES 2005 Conference and Exposition in Africa*, pp. 13–17
2. M.T. Hagh, N. Ghadimi, Multisignal histogram-based islanding detection using neuro-fuzzy algorithm. Complexity **21**(1), 195–205 (2015)
3. M. Geidl, G. Koeppel, P. Favre-Perrod, B. Klockl, G. Andersson, K. Frohlich, Energy hubs for the future. IEEE Power Energy Manag. **5**(1), 24–30 (2007)
4. S. Pazouki, M.-R. Haghifam, Optimal planning and scheduling of energy hub in presence of wind, storage and demand response under uncertainty. Int. J. Electr. Power Energy Syst. **80**, 219–239 (2016)
5. R. Aazami, K. Aflaki, M.R. Haghifam, A demand response based solution for LMP management in power markets. Int. J. Electr. Power Energy Syst. **33**(5), 1125–1132 (2011)
6. A. Ahadi, N. Ghadimi, D. Mirabbasi, An analytical methodology for assessment of smart monitoring impact on future electric power distribution system reliability. Complexity **21**(1), 99–113 (2015)
7. "CPLEX 12." [Online]. Available: https://www.gams.com/latest/docs/S_CPLEX.html. Accessed 15 Jul 2018
8. A. Brooke, D. Kendrick, A. Meeraus, R. Raman, R.E. Rosenthal, *GAMS A User's Guide Introduction 1* (GAMS Development Corporation, Washington, DC, 1998)
9. S. Nojavan, H. allah Aalami, Stochastic energy procurement of large electricity consumer considering photovoltaic, wind-turbine, micro-turbines, energy storage system in the presence of demand response program. Energy Convers. Manag. **103**, 1008–1018 (2015)

10. H.A. Aalami, S. Nojavan, Energy storage system and demand response program effects on stochastic energy procurement of large consumers considering renewable generation. IET Gener. Transm. Distrib. **10**(1), 107–114 (2016)
11. A. Soroudi, M. Aien, M. Ehsan, A probabilistic modeling of photo voltaic modules and wind power generation impact on distribution networks. IEEE Syst. J. **6**(2), 254–259 (2012)
12. F. Jabari, S. Nojavan, B. Mohammadi Ivatloo, M.B.B. Sharifian, Optimal short-term scheduling of a novel tri-generation system in the presence of demand response programs and battery storage system. Energy Convers. Manag. **122**, 95–108 (2016)
13. S. Pazouki, M.-R. Haghifam, A. Moser, Uncertainty modeling in optimal operation of energy hub in presence of wind, storage and demand response. Int. J. Electr. Power Energy Syst. **61**, 335–345 (2014)

Chapter 11
Hybrid Cooling, Heating, and Power Energy Procurement

Hossein Khodaei

Abstract In this chapter, to meet the cooling, heating, and power energy requirement of a large industrial consumer, a novel system is introduced including combined cooling, heating, and power (CCHP) in the combination of renewable energies such as photovoltaic (PV) and wind turbine (WT). In addition, as a heat supplier, a solar thermal storage system (STS) is implemented to meet heating demand of the large consumer. The time-of-use and real-time pricing rates of demand response programs are considered to reduce total energy procurement cost. The problem is solved considering two study cases as without and with implementing STS to investigate the impact of STS on optimization of energy procurement process of a large industrial consumer. The problem is formulated as MILP and solved under CPLEX solver in the GAMS optimization program. The application of the considered system is analyzed through numerical studies.

11.1 Introduction

In this chapter, a new structure is proposed to meet the cooling, heating, and electricity load demand of the large consumer. In order to supply the cooling demand of large consumer, which may be required during the manufacturing process, a combined cooling, heating, and power (CCHP) is utilized in the presence of renewable energies such as wind turbine and photovoltaic systems considering gas and electricity as energy carriers [1, 2]. The developed model is consisted of different forms of energies like gas, power, heat, and cold. The objectives are designed to minimize the operating cost. In order to reduce the operating cost, different types of storage units are implemented to store energy at low-price periods and use the stored energy during high price of load demand periods. Solar thermal storage (STS) [3], electricity storage (ES) [4], ice storage conditioner (ISC) [5], and heat storage (HS) [6] are the most important storage types which are considered in the model. In the developed model, energy carriers as electricity and natural gas are converted to

H. Khodaei (✉)
Department of Engineering, Ardabil Branch, Islamic Azad University, Ardabil, Iran

S. Nojavan et al. (eds.), *Robust Energy Procurement of Large Electricity Consumers*,
https://doi.org/10.1007/978-3-030-03229-6_11

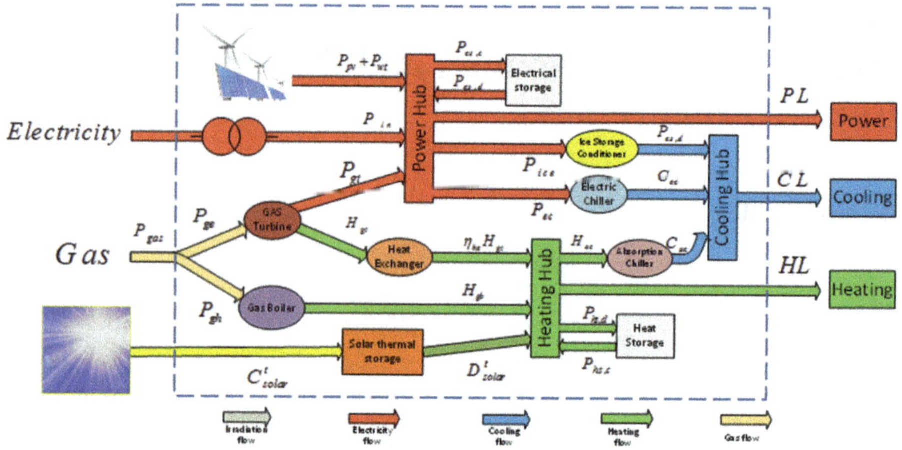

Fig. 11.1 Structure of integrated CCHP for the large consumer

electricity power, cooling, and heat energies considering the gas boiler (GB) [7], gas turbine (GT), absorption chiller (AC) [8], electric chiller (EC) [9], and electrical transformer. The abovementioned devices form the CCHP system, which is used in the developed model for the large consumer. Figure 11.1 illustrates the structure of the model supplying all kinds of demands at the same time. More information about Fig. 11.1 is provided in the following sections.

To show the impact of storage systems, the problem is solved in two cases as with and without STS. The problem is formulated as MILP and solved using the CPLX solver [10] in GAMS optimization software [11].

11.2 Problem Formulation

As said before, the objective function is designed to minimize the operating cost of the developed model. Equation (11.1) models the objective function of the system. Equation (11.2) presents the cost of purchased natural gas from the grid. Also, the cost of the purchased electricity power from the market is provided by Eq. (11.3).

$$\text{Min OF} = M_{\text{pg}} + M_{\text{pe}} \tag{11.1}$$

$$M_{\text{pg}} = P_g \cdot \sum_{t=1}^{24} \left(\frac{P_{\text{gt}}^t}{\eta_{\text{ge}}} + \frac{H_{\text{gb}}^t}{\eta_{\text{gh,gb}}} \right) \cdot \Delta t \tag{11.2}$$

$$M_{\text{pe}} = \sum_{t=1}^{24} \left(P_e^t \cdot P_{\text{in}}^t \cdot \Delta t \right) \tag{11.3}$$

where

M_{pg} is the cost of gas purchased
M_{pe} is the cost of purchased electricity power
P_{gt}^t is the generated power of gas turbine
H_{gb}^t is the output heat of gas boiler
η_{ge} is the efficiency of gas turbine for generating power
$\eta_{\text{gh,gb}}$ is the efficiency of gas boiler for generating heat
P_{in}^t is the actual power of electricity grid

11.2.1 Power Exchange Constraints

In the developed model, the hub is considered in grid-connected model which means it can exchange power with the upstream grid in which excessive generation can be injected to the grid while shortage of power generation is compensated by purchasing power from the grid. Energy exchanging between the electricity network and power hub is presented by Eq. (11.4).

$$P_{\text{grid}}^t \eta_t = P_{\text{in}}^t \tag{11.4}$$

where

P_{grid}^t is the exchange power between electricity grid and energy hub system
η_t is the transformer efficiency

11.2.2 Gas Turbine Model

The gas turbine model is provided in Eq. (11.5). Note that the power output of GT goes to the heat exchanger and power hub:

$$\eta_{\text{ge}} \cdot P_{\text{ge}}^t = P_{\text{gt}}^t \tag{11.5}$$

11.2.3 Limitation of Energy Devices

Every device in multi-energy system has limited capacity. Power generation of each device is limited by (11.6)–(11.10), for photovoltaic, wind turbine, gas turbine, input gas from gas grid, and power exchange between the grid, respectively.

$$0 \leq P_{\mathrm{pv}}^{t} \leq P_{\mathrm{pv}}^{\max} \tag{11.6}$$

$$0 \leq P_{\mathrm{wt}}^{t} \leq P_{\mathrm{wt}}^{\max} \tag{11.7}$$

$$0 \leq P_{\mathrm{gas}}^{t} \leq P_{\mathrm{gas}}^{\max} \tag{11.8}$$

$$0 \leq P_{\mathrm{gt}}^{t} \leq P_{\mathrm{gt}}^{\max} \tag{11.9}$$

$$-P_{\mathrm{grid}}^{\max} \leq P_{\mathrm{grid}}^{t} \leq P_{\mathrm{grid}}^{\max} \tag{11.10}$$

where

$P_{\mathrm{pv}}^{\max}$ is the maximum generated power by photovoltaic system
$P_{\mathrm{wt}}^{\max}$ is the maximum generated power by wind turbine
$P_{\mathrm{gas}}^{\max}$ is the maximum of input gas from gas grid
$P_{gt}^{\max}$ is the maximum generated power from gas turbine
$P_{\mathrm{grid}}^{\max}$ is the maximum exchanged power between electricity grid and energy hub system

11.2.4 Modeling and Constraints of Energy Conversion of Cooling System

The cooling hub receives cooling energy from electric chiller and absorption chiller unilaterally. The absorption and electrical chiller consumes heat and electricity and converts into cold, respectively. Equations (11.11) and (11.12) model the electric and absorption chillers, respectively.

$$P_{\mathrm{ec}}^{t} \cdot \mathrm{COP}_{\mathrm{ec}} = C_{\mathrm{ec}}^{t} \tag{11.11}$$

$$H_{\mathrm{ac}}^{t} \cdot \mathrm{COP}_{\mathrm{ac}} = C_{\mathrm{ac}}^{t} \tag{11.12}$$

where

C_{ec}^{t} is the cooling output of electrical chiller
C_{ac}^{t} is the cooling output of absorption chiller
$\mathrm{COP}_{\mathrm{ec}}$ is the performance coefficient of ice storage conditioner (ISC)
$\mathrm{COP}_{\mathrm{ac}}$ is the coefficient for the performance of absorption chiller
H_{ac}^{t} is the heat input of absorption chiller
P_{ec}^{t} is the input power of chiller

11.2.5 Modeling of the Ice Storage Air Conditioner

The ice storage air conditioner is used as cooling device to meet the cooling demand of the large consumer, which is composed of a chiller and ice storage tank. The chiller is responsible to make ice, which is stored in the ice tank. In the mathematical model (11.13)–(11.15), the cooling energy balance is provided by Eqs. (11.13)–(11.15) limiting the amount of ice storage conditioner and electrical chiller, respectively.

$$P_{\text{ice}}^{t} \cdot \text{COP}_{\text{ice}} = P_{\text{cs},c}^{t} \tag{11.13}$$

$$0 \leq P_{\text{ice}}^{t} \leq P_{\text{ice}}^{\max} \tag{11.14}$$

$$0 \leq P_{\text{ec}}^{t} \leq P_{\text{ec}}^{\max} \tag{11.15}$$

where

COP_{ice} is the performance coefficient of ISC
$P_{\text{cs},c}^{t}$ is the charging cool of ice storage conditioner
P_{ice}^{t} is the input power of the ISC's chiller
$P_{\text{ec}}^{\max}$ is the maximum input power of chiller
$P_{\text{ice}}^{\max}$ is the maximum input power of the ISC's chiller

11.2.6 Modeling of the Heat-Generating Units

As shown in Eq. (11.16), the entrance gas from gas network goes to GB and GT, and then some part of it is converted to electricity and heat. Energy conversion of the gas boiler is presented by Eq. (11.17). Equation (11.18) presents the energy conversion of the gas turbine. Finally, Eqs. (11.19) and (11.20) provide the limitations of heat devices.

$$P_{\text{gas}}^{t} = P_{\text{ge}}^{t} + P_{\text{gh}}^{t} \tag{11.16}$$

$$\eta_{\text{gh,gb}} \cdot P_{\text{gh}}^{t} = H_{\text{gb}}^{t} \tag{11.17}$$

$$\eta_{\text{gh,gt}} \cdot P_{\text{ge}}^{t} = H_{\text{gt}}^{t} \tag{11.18}$$

$$0 \leq H_{\text{ac}}^{t} \leq H_{\text{ac}}^{\max} \tag{11.19}$$

$$0 \leq H_{\text{gb}}^{t} \leq H_{\text{gb}}^{\max} \tag{11.20}$$

11.2.7 Modeling of the Solar Thermal Storage

The STS converts solar irradiation to heat energy and sends it to the heating hub. The mathematical formulation of the STS is provided by Eqs. (11.21)–(11.23). Equations (11.21) and (11.22) limit the charging and discharging of the STS, respectively. Equation (11.23) models the energy loss of the STS, depending on stored energy and ambient temperature and coefficients of STS.

$$C_{\text{solar}}^{t} \leq B_{\text{Ch, solar}}^{t} \times C^{\text{max, Ch, solar}} \times L^{\text{solar}} \tag{11.21}$$

$$D_{\text{solar}}^{t} \leq B_{\text{Dch, solar}}^{t} \times C^{\text{max, Dch, solar}} \times L^{\text{solar}} \tag{11.22}$$

$$P_{\text{loss}}^{t} = S_{\text{solar}}^{t-1} \times \theta^{\text{solar}} + \theta^{\text{static}} \times P_{\text{Unused}}^{t} \tag{11.23}$$

The unused energy of STS is presented in inequality (11.24). Equation (11.25) presents the stored energy in the STS, Eq. (11.26) limits the capacity of the STS, Eq. (11.27) limits the stored energy in the STS, and the energy converted to the heat energy is provided by Eq. (11.28).

$$P_{\text{Unused}}^{t} \geq \frac{\text{Tem}^{\min} - \text{Tem}_{\text{Ambient}}^{t}}{\text{Tem}^{\max} - \text{Tem}^{\min}} \times L^{\text{solar}} \tag{11.24}$$

$$S_{\text{solar}}^{t} = S_{\text{solar}}^{t-1} + \left(\delta \times \eta^{\text{Ch, solar}} \times C_{\text{solar}}^{t}\right) - \sum_{j=1}^{J} \frac{\delta \times D_{\text{solar}}^{t}}{\eta^{\text{Dch, solar}}} - \left(\delta \times P_{\text{loss}}^{t}\right) \tag{11.25}$$

$$S_{\text{solar}}^{t} \leq L^{\text{solar}} \tag{11.26}$$

$$C_{t}^{\text{solar}} \leq Q_{t}^{\text{solar}} \tag{11.27}$$

$$Q_{t}^{\text{solar}} = \varphi_{t} \times A^{\text{solar}} \times \kappa^{\text{solar}} m^{2} \tag{11.28}$$

11.2.8 Energy Storage Model

As said before, in order to reduce the operating cost of the energy hub, different types of energy storages are implemented to store cooling, heating, and electrical energy during the low-price periods and use the stored energy during the high-price periods.

11.2.8.1 Electrical Storage Model

The electrical storage is modeled by Eqs. (11.29)–(11.33). The stored energy in the electrical storage is provided by Eq. (11.29); charging and discharging limitations are presented by Eqs. (11.30) and (11.31), respectively; and stored energy is limited by Eq. (11.32). Constraint (11.33) states that the initial and final amount of the stored energy in the time horizon of the study should be equal.

$$E_{\mathrm{es}}^{t+1} = E_{\mathrm{es}}^{t}(1-\delta_{\mathrm{es}}) + \left(P_{\mathrm{es},c}^{t}\eta_{\mathrm{es},c} - \frac{P_{\mathrm{es},d}^{t}}{\eta_{\mathrm{es},d}}\right)\cdot\Delta t \tag{11.29}$$

$$0 \le P_{\mathrm{es},c}^{t} \le u_{\mathrm{es}}\cdot P_{\mathrm{es},c}^{\max} \tag{11.30}$$

$$0 \le P_{\mathrm{es},d}^{t} \le (1-u_{\mathrm{es}})\cdot P_{\mathrm{es},d}^{\max} \tag{11.31}$$

$$E_{\mathrm{es}}^{\min} \le E_{\mathrm{es}}^{t} \le E_{\mathrm{es}}^{\max} \tag{11.32}$$

$$E_{\mathrm{es}}^{24} = E_{\mathrm{es}}^{0} \tag{11.33}$$

11.2.8.2 Cooling Storage Model

The mathematical model of energy storage of cooling is presented in Eqs. (11.34)–(11.38), which are the same as the electrical storage.

$$E_{\mathrm{cs}}^{t+1} = E_{\mathrm{cs}}^{t}(1-\delta_{\mathrm{cs}}) + \left(P_{\mathrm{cs},c}^{t}\eta_{\mathrm{cs},c} - \frac{P_{\mathrm{cs},d}^{t}}{\eta_{\mathrm{cs},d}}\right)\cdot\Delta t \tag{11.34}$$

$$0 \le P_{\mathrm{cs},c}^{t} \le u_{\mathrm{cs}}\cdot P_{\mathrm{cs},c}^{\max} \tag{11.35}$$

$$0 \le P_{\mathrm{cs},d}^{t} \le (1-u_{\mathrm{cs}})\cdot P_{\mathrm{cs},d}^{\max} \tag{11.36}$$

$$E_{\mathrm{cs}}^{\min} \le E_{\mathrm{cs}}^{t} \le E_{\mathrm{cs}}^{\max} \tag{11.37}$$

$$E_{\mathrm{cs}}^{24} = E_{\mathrm{cs}}^{0} \tag{11.38}$$

11.2.8.3 Heat Storage Model

The heat storage constraints and mathematical model are similar to electrical storage and are expressed in Eqs. (11.39)–(11.43).

$$E_{\mathrm{hs}}^{t+1} = E_{\mathrm{hs}}^{t}(1-\delta_{\mathrm{hs}}) + \left(P_{\mathrm{hs},c}^{t}\eta_{\mathrm{hs},c} - \frac{P_{\mathrm{hs},d}^{t}}{\eta_{\mathrm{hs},d}}\right)\cdot\Delta t \tag{11.39}$$

$$0 \le P_{\mathrm{hs},c}^{t} \le u_{\mathrm{hs}}\cdot P_{\mathrm{hs},c}^{\max} \tag{11.40}$$

$$0 \leq P_{\mathrm{hs},d}^{t} \leq (1 - u_{\mathrm{hs}}) \cdot P_{\mathrm{hs},d}^{\max} \tag{11.41}$$

$$E_{\mathrm{hs}}^{\min} \leq E_{\mathrm{hs}}^{t} \leq E_{\mathrm{hs}}^{\max} \tag{11.42}$$

$$E_{\mathrm{hs}}^{24} = E_{\mathrm{hs}}^{0} \tag{11.43}$$

11.2.9 Real-Time Pricing Demand Response Program

The importance and impact of the real-time pricing are detailed in previous chapters. To reduce operating cost of the energy hub, the real-time pricing (RTP) demand response program (DRP) is implemented. The formulation of the RTP-DRP is provided by Eqs. (11.44)–(11.49). Equation (11.44) presents the aggregation of all power demands in a typical day. Constraints (11.45) and (11.46) express the average power demand and the float factor of RTP, respectively. The mathematical model of the RTP is provided in Eq. (11.47). Equation (11.48) states that the RTP could not exceed from limited amounts. Finally, the mathematical formulation of the DRP is provided in Eq. (11.49).

$$W_d = \sum_{t=1}^{24} \mathrm{PL}^t \tag{11.44}$$

$$P_{\mathrm{av}} = \frac{W_d}{24} \tag{11.45}$$

$$\gamma^t = \frac{\mathrm{PL}^t}{P_{\mathrm{av}}} \tag{11.46}$$

$$\lambda_{\mathrm{RTP}}^{t} = \gamma^t \cdot \lambda_{\mathrm{TOU}}^{t} \tag{11.47}$$

$$\lambda_{\mathrm{RTP}}^{\min} \leq \lambda_{\mathrm{RTP}}^{t} \leq \lambda_{\mathrm{RTP}}^{\max} \tag{11.48}$$

$$\mathrm{PL}_{\mathrm{RTP}}^{t} = \mathrm{PL}^t + E \cdot \mathrm{PL}^t \left(\frac{\lambda_{\mathrm{RTP}}^{t} - \lambda_{\mathrm{TOU}}^{t}}{\lambda_{\mathrm{TOU}}^{t}} \right) \tag{11.49}$$

11.2.10 The Balance of Energy Hub System

The energy input and output of energy hub should be equal. Equation (11.50) shows the energy balance of the power hub, which the left side represents input power and the right side is the output power. For cooling hub, Eq. (11.51) expresses that cooling demand is supplied by absorption chiller, electric chiller, and cooling of ISC discharging which are in the left side. Finally, the heating balance of heating hub is presented in Eq. (11.52).

$$P_{\text{in}}^{t} + P_{\text{pv}}^{t} + P_{\text{wt}}^{t} + P_{\text{gt}}^{t} + P_{\text{es},d}^{t} = P_{\text{ec}}^{t} + P_{\text{es},c}^{t} + P_{\text{ice}}^{t} + PL_{\text{DRP}}^{t} \tag{11.50}$$

$$C_{\text{ac}}^{t} + C_{\text{ec}}^{t} + P_{\text{cs},d}^{t} = \text{CL}^{t} \tag{11.51}$$

$$\eta_{\text{he}} H_{\text{gt}}^{t} + H_{\text{gb}}^{t} + P_{\text{hs},d}^{t} + D_{\text{solar}}^{t} = H_{\text{ac}}^{t} + P_{\text{hs},c}^{t} + \text{HL}^{t} \tag{11.52}$$

11.3 Case Study

In this section, required parameters and information are provided, and analysis of the optimal obtained solution by consideration of DRP in two cases of with and without STS is studied. The effects of implementing solar thermal storage in energy hub system are investigated.

11.3.1 The Input Data of Energy Hub

The parameters of energy carriers and devices are presented in this section, and all the analysis of energy hub system is based on the input data. The power output of the PV and WT systems is depicted in Fig. 11.2. The energy demand of power, cooling, and heating in a typical summer day is shown in Fig. 11.3.

The CCHP, WT, and PV coefficients are presented in Table 11.1. The parameters of storage devices are presented in Table 11.2. Table 11.3 presents the parameters of STS. Finally, the gas and electricity prices are presented in Table 11.4.

11.3.2 Simulation Results

The problem is solved in two cases to prove the effectiveness and advantage of the developed model. The considered case studies are as follows:

Case 1: Optimal scheduling of the energy hub system without STS
Case 2: Optimal scheduling of the energy hub system in the presence of STS

Based on the obtained results, total operating cost of the energy hub is equal to \$7787.552 and \$7489.336 under RTP-DRP for without and with implementing the STS. According to obtained results by implementing STS, the operation cost is reduced by 3.83% which indicates a significant development in economic aspect of energy hub system. Comparison results of two cases are presented in Table 11.5.

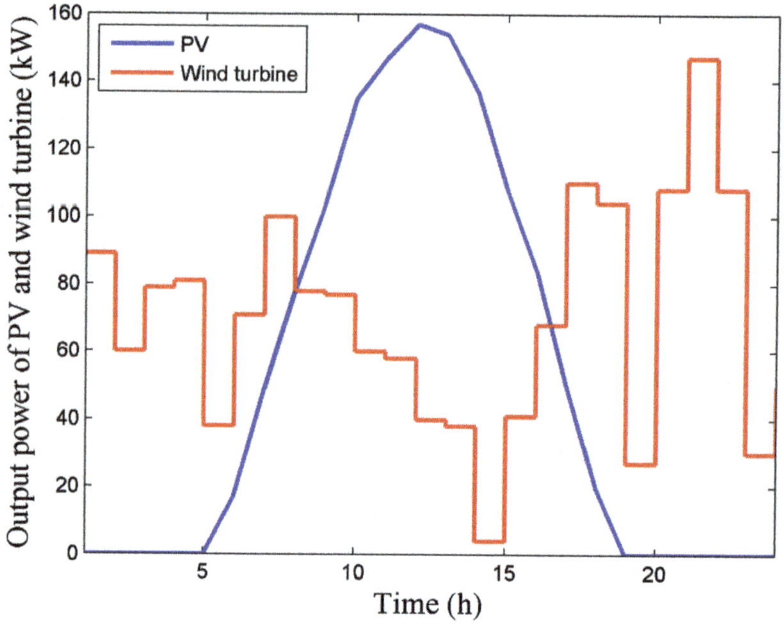

Fig. 11.2 The power output of WT and PV systems

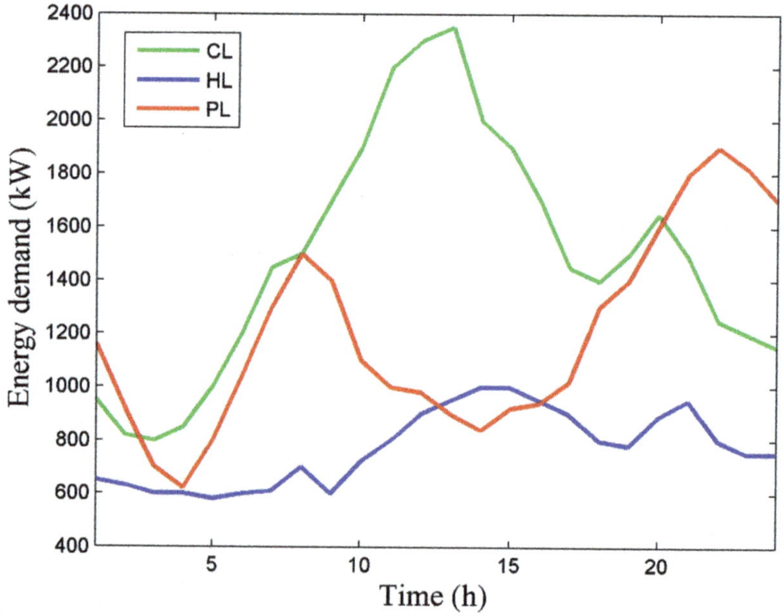

Fig. 11.3 The power, cooling, and heating demands of the micro grid in a typical summer day

Table 11.1 Parameters of coefficients of CCHP, WT, PV, and carbon emission

Parameter	Unit	Value	Parameter	Unit	Value	Parameter	Unit	Value
P_{gt}^{max}	kw	1000	P_{grid}^{max}	kw	1500	COP_{ec}	–	4
H_{gb}^{max}	kw	800	P_{gas}^{max}	kw	3400	COP_{ice}	–	3.5
H_{ac}^{max}	kw	1000	η_{ge}	–	0.3	COP_{ac}	–	1.2
P_{ec}^{max}	kw	500	$\eta_{gh,gb}$	–	0.9	β_e	Kg/kwh	0.972
P_{ice}^{max}	kw	100	$\eta_{gh,gt}$	–	0.4	β_g	Kg/kwh	0.23
P_{pv}^{max}	kw	180	η_t	–	0.98	E	–	−0.5
P_{wt}^{max}	kw	200	η_{he}	–	0.7			

Table 11.2 Parameters of storage devices

Electrical storage parameters			Heat storage parameters			Cool storage parameter		
Parameter	Unit	Value	Parameter	Unit	Value	Parameter	Unit	Value
$P_{es,c}^{max}$	kw	500	$P_{hs,c}^{max}$	kW	800	$P_{cs,c}^{max}$	kW	700
$P_{es,d}^{max}$	kw	700	$P_{hs,d}^{max}$	kW	800	$P_{cs,d}^{max}$	kW	800
E_{es}^{min}	kwh	400	E_{hs}^{min}	kWh	400	E_{cs}^{min}	kWh	400
E_{es}^{max}	kwh	1800	E_{hs}^{max}	kWh	1800	E_{cs}^{max}	kWh	1800
$\eta_{es,c}$	–	0.96	$\eta_{hs,c}$	–	0.98	$\eta_{cs,c}$	–	0.97
$\eta_{es,d}$	–	0.96	$\eta_{hs,d}$	–	0.98	$\eta_{cs,d}$	–	0.95
δ_{es}	–	0.01	δ_{hs}	–	0.02	δ_{cs}	–	0.02

Table 11.3 Parameters of STS

Solar thermal storage					
Parameter	Unit	Value	Parameter	Unit	Value
$\eta^{Ch,solar}$	–	0.9	$\eta^{Dch,solar}$	–	0.9
L^{solar}	kwh	1000	Tem^{max}	°C	65
Tem^{min}	°C	36	$C^{max,Ch,solar}$	–	0.3
$C^{max,Dch,solar}$	–	0.3	θ^{static}	–	0.056
θ^{solar}	–	0.057	κ^{solar}	–	0.95
A^{solar}	m^2	500			

Table 11.4 Parameters of gas and electricity prices

Parameter	Unit	Value
λ_{RTP}^{max}	Dollar	47
λ_{RTP}^{min}	Dollar	8
P_g	Dollar/kwh	10.5

Table 11.5 Comparison results of the two cases

Parameters	Case 1	Case 2
Total cost ($)	7787.552	7489.336
Cost reduction (%)	0	3.83

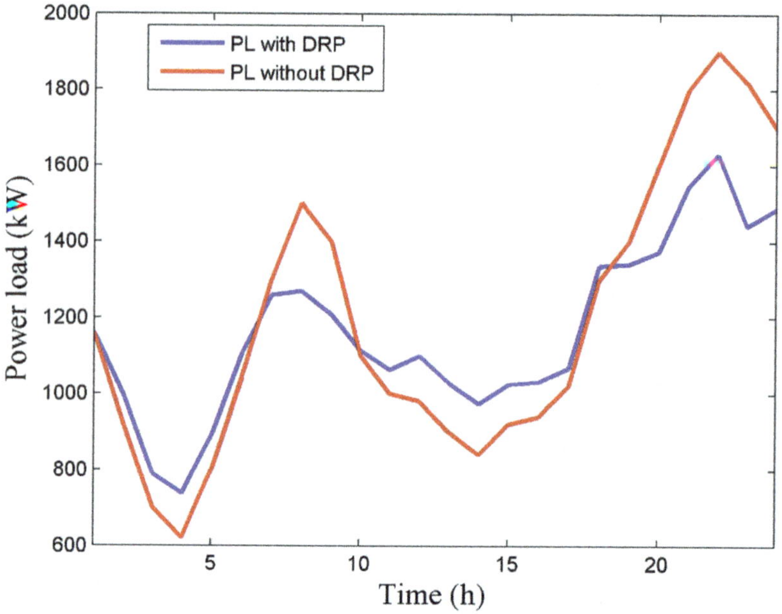

Fig. 11.4 Power load in with and without DRP

11.3.2.1 Load Demand Under DRP

By implementing the DRP, some amounts of loads are shifted from peak to off-peak periods which flattens the demand curve. The electrical load demand is illustrated in Fig. 11.4 for with and without DRP cases. According to Fig. 11.4, the maximum load demand is equal to 1.9 MW at hour 22 in without DRP case, which is reduced to 1.63 MW after applying RTP rate of DRP. In addition, the minimum power load (PL) demand in without DRP case is 0.620 MW at hour 4 which is increased to 0.736 in with DRP case.

As said before, in order to calculate the real-time power prices in RTP demand response program, the information of the time-of-use pricing is required. The RTP pricing against the time-of-use pricing is depicted in Fig. 11.5.

The ambient temperature and solar irradiation are depicted in Fig. 11.6 for each hour in the time horizon of the study. It is obvious that power generation of the PV is maximum when the sun reaches its climax at hour 12.

11.3.2.2 STS Storage

Stored energy, charging and discharging, and energy loss of the STS are depicted in Fig. 11.7. Note that the STS may charge and discharge simultaneously. The maximum charged energy is 300 kW, recorded at hours 9–14, while the STS is not

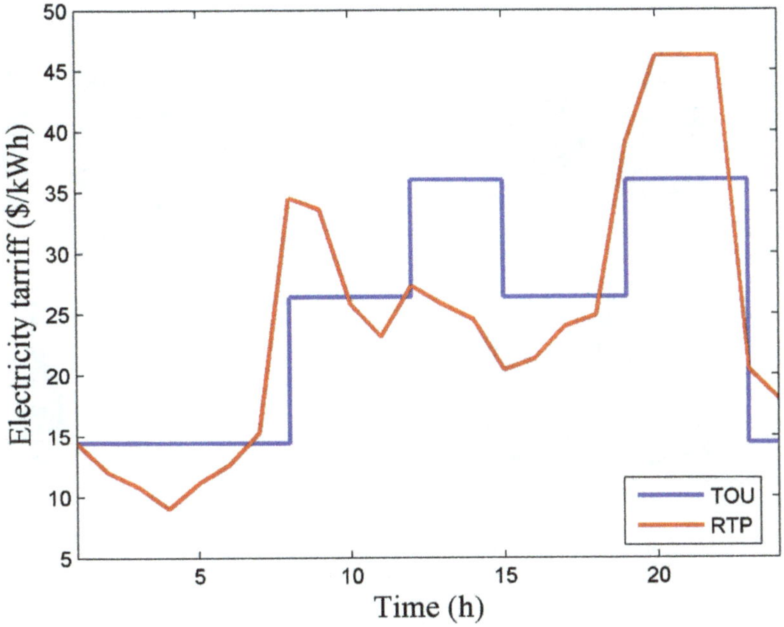

Fig. 11.5 Electricity tariffs for TOU and RTP demand response programs

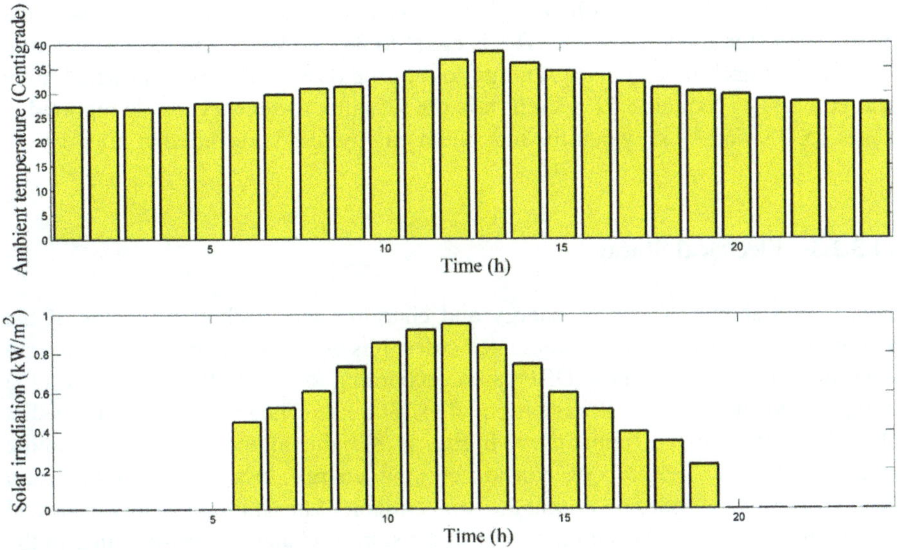

Fig. 11.6 Ambient temperature and solar irradiation

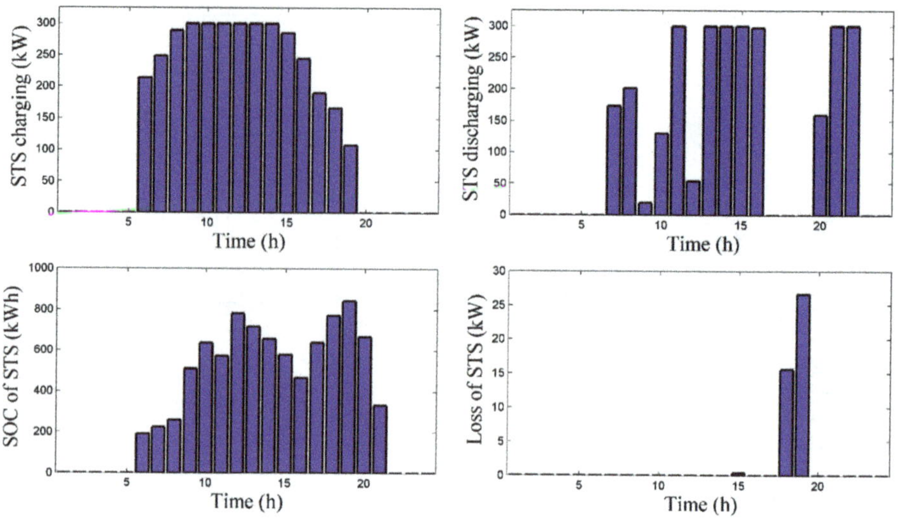

Fig. 11.7 Stored energy and charge and discharging of the STS

charged at hours 0–5 and 20–24. It should be noted that the power output of the PV is zero at this period.

The maximum discharged energy is equal to 300 kW recorded at hours 11, 13–16, and 21–22. It is obvious that the STS is charge and discharged at hours 6–16. Total stored energy in the STS is equal to 8851.0 kWh, which is depicted in Fig. 11.7 for each hour. Finally, energy loss of the STS is illustrated in which total wasted energy is equal to 42.7 kWh. It is obvious that the energy loss is reported at hours 15, 18, and 19 in which the maximum energy loss is recorded at hour 19.

11.3.2.3 Electrical Storage

Figure 11.8 shows the stored energy and charging and discharging states of the storage system for with and without DRP cases. It is obvious that obtained results are the same for with and without DRP cases. According to Fig. 11.8 (lower curve), the initial stored energy in the storage is equal to 1800 kW. The storage is discharged at about 700 kW at hour 1 (upper curve in Fig. 11.8) reducing the stored energy in the storage equal to 1052.8 kW considering the energy loss. The charging and discharging process goes on during the time horizon of the study to supply electrical load demand in which the stored energy in the storage changes corresponding to the charged or discharge power.

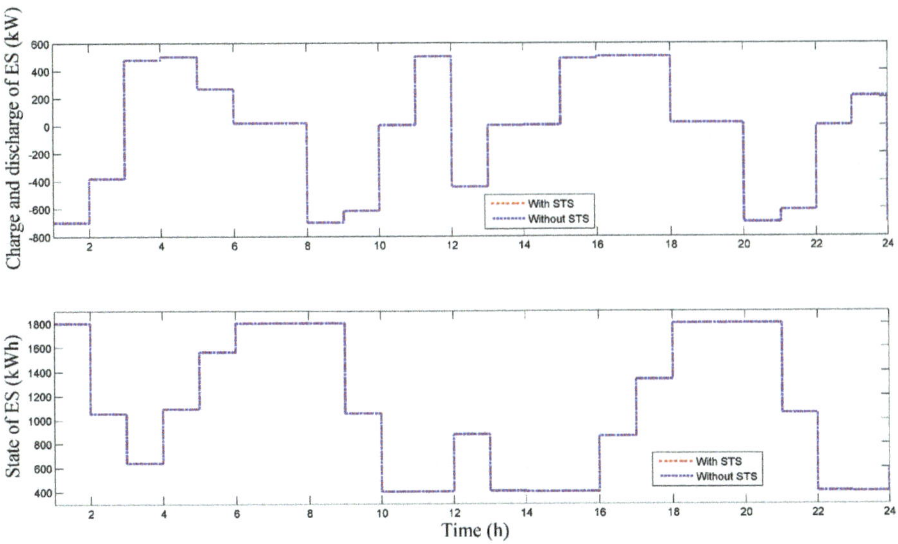

Fig. 11.8 Stored energy and charge and discharging of the electrical storage

11.3.2.4 Thermal Storage

The stored thermal energy and charging and discharging states of the thermal storage system are depicted in Fig. 11.9 for with and without DRP cases. The initial stored thermal energy in the heat storage is equal to 1800 kW. At hour 1, the storage is discharged at about 650 kW (upper curve in Fig. 11.9) reducing the stored energy in the storage equal to 1100.5 kW considering the energy loss. The charging and discharging of the heat storage is repeated during the 24 h to meet the thermal demand of the energy hub. According to the lower curve in Fig. 11.9, stored thermal energy in without STS case is equal to 18,749 kW, which is reduced to 16,673 kW in with STS case indicating the impact of STS on the performance of the thermal storage.

11.3.2.5 Cooling Storage

The stored energy and charging and discharging states of the cooling storage system are depicted in Fig. 11.10. The description of the first hour is almost the same with the previous cases because the modeling of the storage is similar to each other. Obtained results for cooling storage are almost equal for with and without DRP cases in which total stored cooling energy in the storage is equal to 26,330 and 25,882 kW for with and without STS cases, respectively.

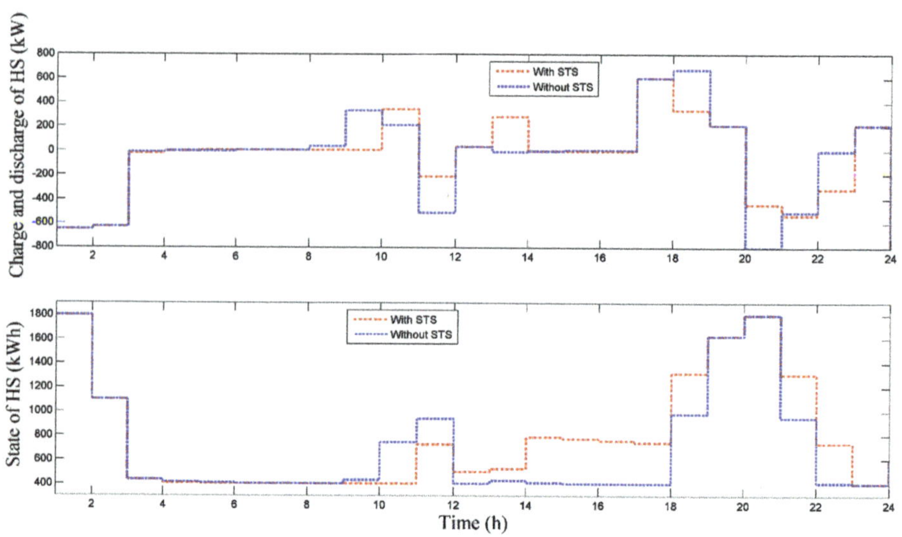

Fig. 11.9 Stored energy and charge and discharging of the thermal storage

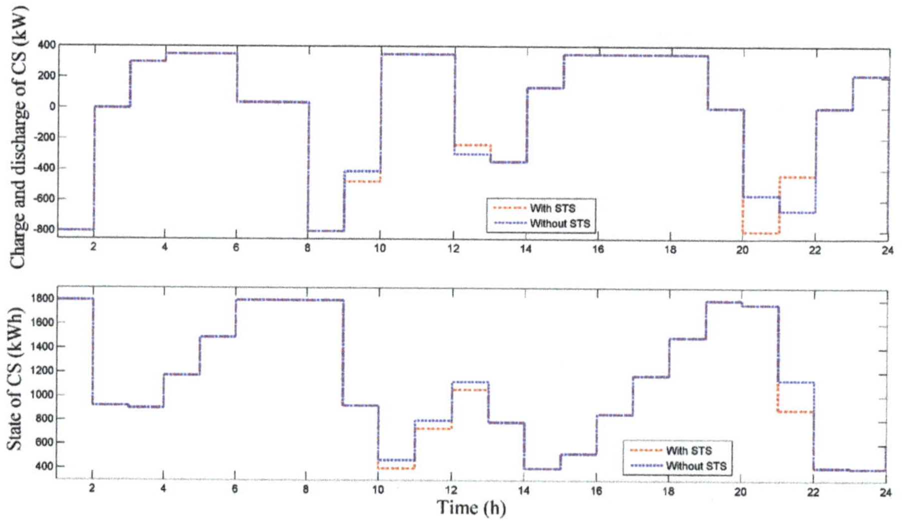

Fig. 11.10 Stored energy and charge and discharging of the cooling storage

11.3.2.6 Cooling Load Demand

Figure 11.11 illustrates the cooling load (CL) against the different cooling generating units at each hour for with and without STS cases. The load demand is presented by a negative value, while positive values present the cooling generation. As it can be seen from Fig. 11.11 and aforementioned descriptions, absorption chiller (AC),

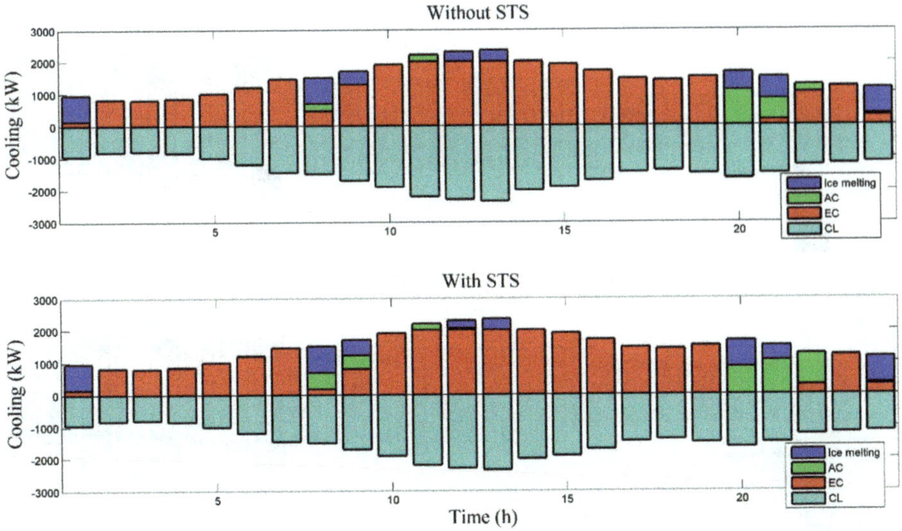

Fig. 11.11 Cooling load demand against the generation of the cooling units

electric chiller (EC), and ice melting supply the cooling load. For example, at hour 21 in without STS case, the cooling load is 1490 kW. The ice melting and AC units provide 667 and 664 kW of the required load, respectively. The rest of required load which is equal to 159 kW is supplied by EC.

In with STS case, the required load at hour 21 is supplied by ice melting and AC systems, in which AC generates 1050 kW and ice melting generated the remainder part which is equal to 440 kW. The aggregation of them is equal to 1490 kW which is the cooling demand at this hour. Note that the same description can be presented for each hour.

11.3.2.7 Heat Load Demand

The required thermal demand against the different thermal units generating is illustrated in Fig. 11.12 at each hour for with and without STS cases. The load demand is presented by negative value, while positive values present the cooling generation. As it can be seen, the thermal demand is composed of heat demand (HL), charging power of heat storage (HS), and required heat for AC units.

In without STS case, the thermal demand at hour 21 is equal to 1500 kW in which 950 kW is the heat load and the rest of demand is required by AC. This amount of demand is supplied by GT and HS discharging providing 993 and 507 kW, respectively.

In with STS case, the thermal demand is equal to 1830 kW at hour 21 in which 950 kW is the heat load and AC requires 880 kW. The thermal load demand is supplied by GT, HS discharging, and STS providing 933, 537, and 300 kW, respectively.

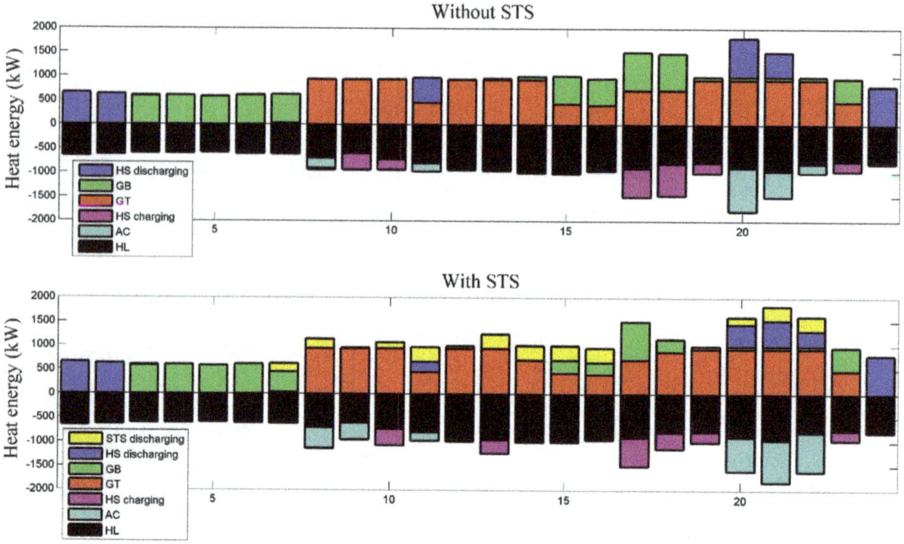

Fig. 11.12 Heating load demand against the generating units

11.3.2.8 Electric Load Demand

The electric demand of the energy hub against the different thermal units generating is depicted in Fig. 11.13 for with and without STS cases. Like the previous figures, the load demand is presented by a negative value, while positive values present the cooling generation. In this case, the electrical demand includes electric power load (PL), electric chiller (EC), charging the electric storage, sold power to the grid, and ISC. Electric-generating units are wind farm, photovoltaic system, GT, ES discharge, and purchased power from the grid.

In without STS case, the electric demand at hour 16 is equal to 1970 kW including 940 kW for PL, 430 kW for EC, 500 for charging the ES, and 100 kW for ISC. The GT, upstream grid, PV, and WT units generate 434, 1466, 90, and 70 kW, to supply the load demand, respectively. Note that generated power is higher than required demand because of energy loss during generating process. The same results are obtained for with using the STS.

11.4 Conclusion

In this chapter, a novel model of energy hub system is developed for large consumer to supply the electric, cooling, and heating load demands. The sub energy integrated with combined cooling, heating, and power (CCHP) is the proposed model. Also, storage devices such as STS and renewable energies are implemented to increase the flexibility and efficiency of the system. In addition, to reduce the operating cost of

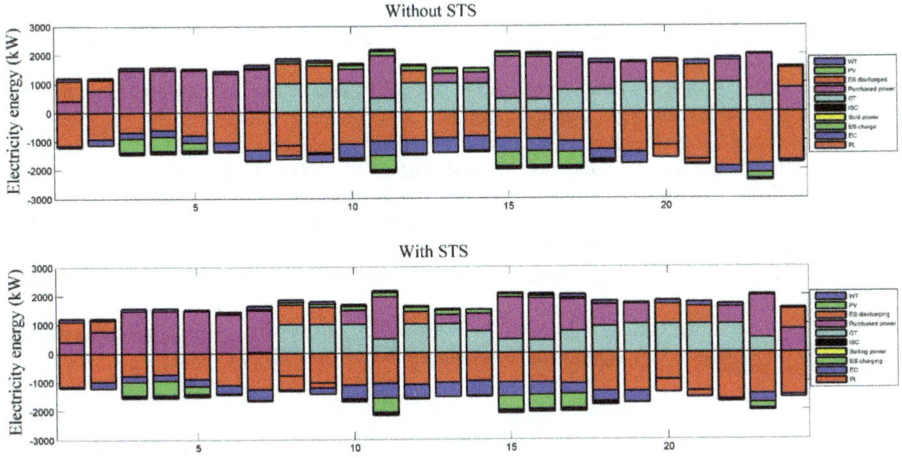

Fig. 11.13 Electric load demand against the generating units

the energy hub, real-time pricing demand response program is utilized. The problem is formulated as mixed-integer linear programming and solved using the CPLX solver in the GAMS optimization software. Two case studies are investigated including with and without STS, and the results are presented to show the impact of STS. In case one which is without STS, the operation cost is equal to \$7787.552. In the second case, when STS is employed, the operation cost obtained \$7489.336. Comparison of two case results shows that implementing STS reduces operation cost by 3.83%. Therefore, by employing STS, significant economic development is achieved in the proposed model. Finally, the performance of different components of the energy hub is investigated.

References

1. F. Jabari, S. Nojavan, B. Mohammadi Ivatloo, Designing and optimizing a novel advanced adiabatic compressed air energy storage and air source heat pump based μ-combined cooling, heating and power system. Energy **116**, 64–77 (2016)
2. M.H. Firouz, N. Ghadimi, Concordant controllers based on FACTS and FPSS for solving wide-area in multi-machine power system. J. Intell. Fuzzy Syst. **30**(2), 845–859 (2016)
3. I. Llorente García, J.L. Álvarez, D. Blanco, Performance model for parabolic trough solar thermal power plants with thermal storage: comparison to operating plant data. Sol. Energy **85** (10), 2443–2460 (2011)
4. S. Nojavan, H. allah Aalami, Stochastic energy procurement of large electricity consumer considering photovoltaic, wind-turbine, micro-turbines, energy storage system in the presence of demand response program. Energy Convers. Manag. **103**, 1008–1018 (2015)
5. G. Fang, X. Liu, S. Wu, Experimental investigation on performance of ice storage air-conditioning system with separate heat pipe. Exp. Thermal Fluid Sci. **33**(8), 1149–1155 (2009)

6. A.M. Khudhair, M.M. Farid, A review on energy conservation in building applications with thermal storage by latent heat using phase change materials. Energy Convers. Manag. **45**(2), 263–275 (2004)
7. K. Zhu, J. Xia, X. Xie, Y. Jiang, Total heat recovery of gas boiler by absorption heat pump and direct-contact heat exchanger. Appl. Therm. Eng. **71**(1), 213–218 (2014)
8. A. Lazrak et al., Development of a dynamic artificial neural network model of an absorption chiller and its experimental validation. Renew. Energy **86**, 1009–1022 (2016)
9. S.F. Afzali, V. Mahalec, Optimal design, operation and analytical criteria for determining optimal operating modes of a CCHP with fired HRSG, boiler, electric chiller and absorption chiller. Energy **139**, 1052–1065 (2017)
10. CPLEX 12. [Online]. Available: https://www.gams.com/latest/docs/S_CPLEX.html. Accessed 15 July 2018
11. A. Brooke, D. Kendrick, A. Meeraus, R. Raman, R.E. Rosenthal, *GAMS a User's Guide Introduction 1* (GAMS Development Corporation, Washington, DC, 1998)

Chapter 12
A Cost- and Environmental-Based Energy Procurement

Mehrdad Khaksar

Abstract In the recent years, considering environmental issues, generated emission by power-generating units has turned to be a challenge in the power system industry. However, financial issues have been always the first priority in power generation and procurement process, which conflicts with environmental objects. Therefore, a trade-off should be made between financial and environmental objects. In this chapter, a multi-objective model is presented to solve the power procurement problem of a large consumer, to take both environmental and economic objectives into account. The ε-constraint and max-min fuzzy satisfying methods are employed to solve and select the trade-off solution, respectively. In addition, to reduce power procurement cost, demand response programs are implemented flattening the load demand curve of the large consumer. The model is formulated as a mixed-integer linear programming and solved using GAMS optimization software. Considered sample of large consumer in Chap. 2 is used to validate efficiency of utilized techniques.

12.1 Introduction

Considering environmental issues such as climate change and greenhouse gas emission has drawn attentions to supply required energy of the consumer in an environment-friendly way especially for large consumers, which consume a great deal of energy. To deal with the financial and environmental aspects of energy generation or procurement processes, which are conflicting with each other [1], different methods as the weighed sum [2], ε-constraint [3], goal programming [4], compromise programming [5], etc. are utilized in the literature.

In order to supply different types of required energy in a manufacturing process by a large consumer considering financial and environmental issues, the energy hub systems are proposed as a main solution. Therefore, in this chapter, an energy hub system including different types of energy resources with various types of loads as

M. Khaksar (✉)
Young Researchers and Elite Club, Islamshahr Branch, Islamic Azad University, Islamshahr, Iran

S. Nojavan et al. (eds.), *Robust Energy Procurement of Large Electricity Consumers*,
https://doi.org/10.1007/978-3-030-03229-6_12

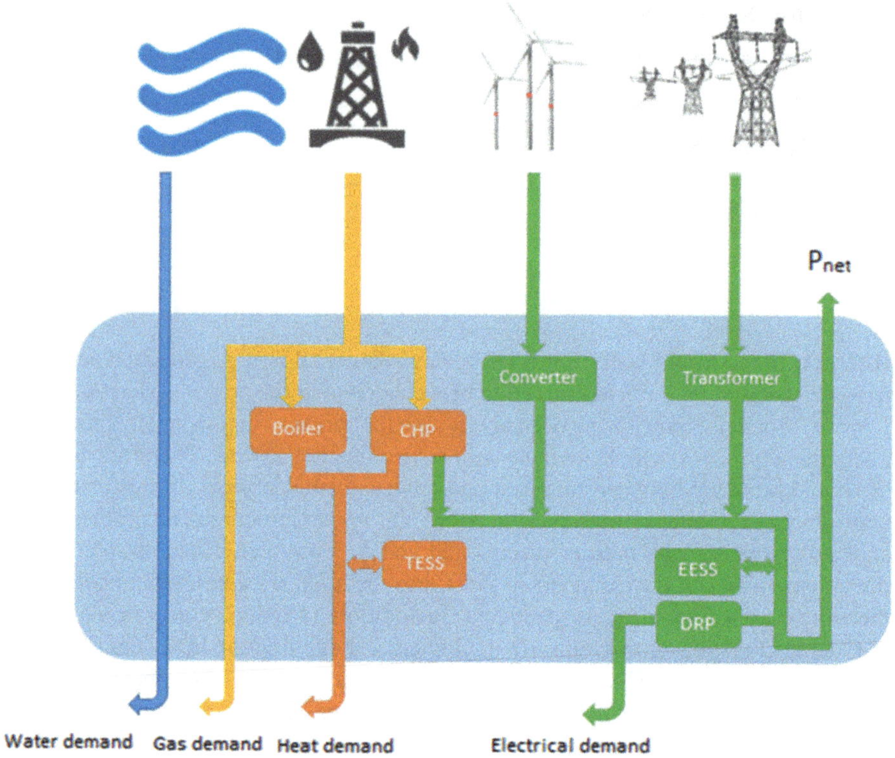

Fig. 12.1 Hub energy system

illustrated in Fig. 12.1 is utilized to solve the energy procurement problem of a large consumer considering financial and environmental challenges.

According to Fig. 12.1, renewable generation units, CHP unit, electrical storages, and the upstream grid supply the electrical load demand under the time-of-use rate of demand response program. Purchased gas from the grid is used for various purposes as supplying residential gas demand, feeding the boiler and CHP units. In addition to the abovementioned energy sources, in order to save energy during the excess production periods, the thermal energy storage system is used in the hub system. In addition, in order to flatten the load profile, which results in reduction of the total operating cost and emission, the time-of-use rate (TOU) of demand response program (DRP) has been implemented.

12.2 Mathematical Formulation

Improving the economic and environmental operation of a large consumer is the main goal of the developed multi-objective energy hub model [6]. In the following subsections, the mathematical formulation of the optimal economic and environmental scheduling problem of hub energy system under TOU-DRP is provided in the following subsections.

12.2.1 The First Objective Function

The operating cost of the system is taken as the first objective function of the developed multi-objective model, which is presented by Eq. (12.1). The aim is to minimize the first objective function [7]:

$$\mathrm{Min} f_1 = \mathrm{Cost} = \sum_{t}^{H} (B + C + D + E + F + G + H + I + J) \tag{12.1}$$

$$B = \lambda_t^e \times p_t^e \tag{12.2}$$

$$C = \lambda^{\mathrm{wi}} \times p_t^{\mathrm{wi}} \tag{12.3}$$

$$D = \lambda_s^e \times \left(p_t^{\mathrm{ch},e} + p_t^{\mathrm{dis},e}\right) \tag{12.4}$$

$$E = \lambda^{\mathrm{DR}} \times \left(p_t^{e,\mathrm{shdo}} + p_t^{e,\mathrm{shup}}\right) \tag{12.5}$$

$$F = \lambda_t^e \times \left(p_t^{\mathrm{ch},e} - p_t^{\mathrm{dis},e}\right) \tag{12.6}$$

$$G = \lambda^g \times g_t^{\mathrm{CHP}} \tag{12.7}$$

$$H = \lambda^g \times g_t^B \tag{12.8}$$

$$I = \lambda_s^h \times \left(p_t^{\mathrm{ch},h} + p_t^{\mathrm{dis},h}\right) \tag{12.9}$$

$$J = \lambda^{\mathrm{wa}} \times \mathrm{wa}_t \tag{12.10}$$

where Cost is the total operating cost; p_t^e is the purchased power from the market at price λ_t^e; p_t^{wi} is the produced power from the wind turbine at price λ^{wi}; $p_t^{\mathrm{ch},e}$ and $p_t^{\mathrm{dis},e}$ are the charged and discharged power of the electrical storage, respectively, at operating price λ_s^e; $p_t^{e,\mathrm{shdo}}$ is the reduced electrical load; $p_t^{e,\mathrm{shup}}$ is the increased electrical load; λ^{DR} is the cost of demand response; g_t^{CHP} is the gas consumption of CHP; g_t^B is the gas consumption of boiler; λ^g is the gas price; $p_t^{\mathrm{ch},h}$ is the charged heat of the thermal storage; $p_t^{\mathrm{dis},h}$ is the discharged heat of thermal storage; λ_s^h is the operating cost of heat storage; and wa_t is the purchased water from the grid at price λ^{wa}.

In Eqs (12.2)–(12.10), the cost of purchased power from the grid is presented by Eq. (12.2), cost of electricity generation of wind turbine by Eq. (12.3), operating cost of electrical storage by Eq. (12.4), cost of DRP implementation by Eq. (12.5), cost or revenue of exchanged power by Eq. (12.6), operating cost of CHP by Eq. (12.7), operating cost of the boiler by Eq. (12.8), operating cost of thermal storage by Eq. (12.9), and finally, cost of purchased water for supplying water demand by Eq. (12.10).

12.2.2 The Second Objective Function

Produced emission during the power- and heat-generating process consisting SO2, NO2, and CO2 is taken as the second objective function in the developed bi-objective model presented by Eq. (12.11). The aim is to minimize Eq. (12.11):

$$\mathrm{Min} f_2 = \mathrm{Em} = \left(\mathrm{Em}^{\mathrm{CHP}} + \mathrm{Em}^{B} + \mathrm{Em}^{L} + \mathrm{Em}^{\mathrm{NET}}\right) \tag{12.11}$$

$$\mathrm{Em}^{\mathrm{CHP}} = \left(\mathrm{EF}_{\mathrm{CO}}^{\mathrm{CHP}} \times g_t^{\mathrm{CHP}}\right) + \left(\mathrm{EF}_{\mathrm{SO}}^{\mathrm{CHP}} \times g_t^{\mathrm{CHP}}\right) + \left(\mathrm{EF}_{\mathrm{NO}}^{\mathrm{CHP}} \times g_t^{\mathrm{CHP}}\right) \tag{12.12}$$

$$\mathrm{Em}^{B} = \left(\mathrm{EF}_{\mathrm{CO}}^{B} \times g_t^{B}\right) + \left(\mathrm{EF}_{\mathrm{SO}}^{B} \times g_t^{B}\right) + \left(\mathrm{EF}_{\mathrm{NO}}^{B} \times g_t^{B}\right) \tag{12.13}$$

$$\mathrm{Em}^{L} = \left(\mathrm{EF}_{\mathrm{CO}}^{L} \times g_t^{L}\right) + \left(\mathrm{EF}_{\mathrm{SO}}^{L} \times g_t^{L}\right) + \left(\mathrm{EF}_{\mathrm{NO}}^{L} \times g_t^{L}\right) \tag{12.14}$$

$$\mathrm{Em}^{\mathrm{NET}} = \left(\mathrm{EF}_{\mathrm{CO}}^{\mathrm{NET}} \times p_t^{e}\right) + \left(\mathrm{EF}_{\mathrm{SO}}^{\mathrm{NET}} \times p_t^{e}\right) + \left(\mathrm{EF}_{\mathrm{NO}}^{\mathrm{NET}} \times p_t^{e}\right) \tag{12.15}$$

where Em is the total emission of energy hub system, $\mathrm{Em}^{\mathrm{CHP}}$ is the produced emission of CHP unit, Em^{B} is the produced emission by the boiler, $\mathrm{Em}^{\mathrm{NET}}$ is the emission of purchased power from the market, and Em^{L} is the emission of consumed residential gas. Note that each of Eqs. (12.12)–(12.15) presents produced emission by the CHP, boiler, residential consumer, and power generation in power plants, respectively.

12.2.3 Operating Requirements of Thermal Units

In order to supply heat demand of the energy hub system, various sources such as boiler, CHP unit, and thermal storage unit systems are utilized. Equation (12.16) provides balance between heat demand and generated heat by thermal units:

$$p_t^{h} = \left[\eta_{\mathrm{gh}}^{B} \times g_t^{B}\right] + \left[A^{\mathrm{CHP}} \times \eta_{\mathrm{gh}}^{\mathrm{CHP}} \times g_t^{\mathrm{CHP}}\right] + \left(p_t^{\mathrm{dis},h} - p_t^{\mathrm{ch},h}\right) \tag{12.16}$$

where η_{gh}^{B} is the gas to heat efficiency of the boiler, A^{CHP} is the availability of CHP unit, and $\eta_{\mathrm{ge}}^{\mathrm{CHP}}$ is the gas to electricity efficiency of CHP.

The thermal storage unit is implemented to store excessive generated heat in which the available heat in the thermal storage is presented by Eq. (12.17):

$$C_t^{\mathrm{st},h} = C_{t-1}^{\mathrm{st},h} + p_t^{\mathrm{ch},h} \times \eta_{\mathrm{ch}}^{h} - p_t^{\mathrm{dis},h} / \eta_{\mathrm{dis}}^{h} - p_t^{\mathrm{loss},h} \tag{12.17}$$

where $C_t^{\mathrm{st},h}$ is the state of charge of thermal storage and $p_t^{\mathrm{loss},h}$ is the heat loss of thermal storage.

Equation (12.18) models the heat loss as a function of available energy $\alpha_{\mathrm{loss}}^{h}$ in the thermal storage system. Equation (12.19) limits the stored heat energy in the storage:

$$p_t^{\text{loss},h} = \alpha_{\text{loss}}^h \times C_t^{\text{st},h} \tag{12.18}$$

$$\alpha_{\min}^h \times C_c^{\text{st},h} \le C_t^{\text{st},h} \le \alpha_{\max}^h \times C_c^{\text{st},h} \tag{12.19}$$

where $\alpha_{\min}^h$ and $\alpha_{\max}^h$ are coefficients for modeling the minimum and maximum capacity of the thermal storage, respectively.

Charging and discharging heat energy of storage is limited by using Eqs. (12.20) and (12.21), respectively:

$$\frac{\alpha_{\min}^h \times C_c^{\text{st},h} \times I_t^{\text{ch},h}}{\eta_{\text{ch}}^h} \le p_t^{\text{ch},h} \le \frac{\alpha_{\max}^h \times C_c^{\text{st},h} \times I_t^{\text{ch},h}}{\eta_{\text{ch}}^h} \tag{12.20}$$

$$\alpha_{\min}^h \times C_c^{\text{st},h} \times I_t^{\text{dis},h} \times \eta_{\text{dis}}^h \le p_t^{\text{dis},h} \le \alpha_{\max}^h \times C_c^{\text{st},h} \times I_t^{\text{dis},h} \times \eta_{\text{dis}}^h \tag{12.21}$$

where $I_t^{\text{ch},h}/I_t^{\text{dis},h}$ is binary variable to model charging/discharging state of thermal storage. The charging and discharging at the same time is prevented using Eq. (12.22):

$$I_t^{\text{ch},h} + I_t^{\text{dis},h} \le 1 \tag{12.22}$$

12.2.4 Operating Requirements of Electrical Units

The electrical demand after applying DRP is supplied procuring power from different power procurement sources such as the upstream grid, wind turbine, CHP unit, and electrical storage system. The balance between procured electrical power and load demand is stated by Eq. (12.23):

$$\begin{aligned} p_t^l + p_t^{\text{shup}} - p_t^{\text{shdo}} = {} & \left[A^{\text{NET}} \times \eta_{\text{ee}}^T \times p_t^e\right] + \left[A^{\text{WIND}} \times \eta_{\text{ee}}^{\text{CON}} \times p_t^{\text{wi}}\right] \\ & + \left[A^{\text{CHP}} \times \eta_{\text{ge}}^{\text{CHP}} \times g_t^{\text{CHP}}\right] + \left[p_t^{\text{dis},e} - p_t^{\text{ch},e}\right] \end{aligned} \tag{12.23}$$

where A^{NET} is the availability of the upstream grid, A^{WIND} is the availability of the wind turbine, and $\eta_{\text{ee}}^{\text{CON}}$ is the efficiency of the converter.

In this chapter, it is assumed that the purchased power from the upstream grid cannot exceed a rated capacity of transformer, which is declared by Eq. (12.24):

$$\eta_{\text{ee}}^T \times p_t^e \le p_c^T \tag{12.24}$$

where η_{ee}^T is the efficiency of the electricity transformer and p_c^T is the rated capacity of the transformer.

The wind turbine is considered as the only renewable energy source which is modeled by Eq. (12.25):

$$p_t^{\text{wi}} = \begin{cases} 0 & w < w_{\text{ci}} \\ p_r(z - y.w(t) + x.w^2(t)) & w_{\text{ci}} \leq w < w_r \\ p_r & w_r \leq w < w_{\text{co}} \\ 0 & w \geq w_{\text{co}} \end{cases} \tag{12.25}$$

where:

p_r is the rated power of wind turbine.
x, y, and z are coefficients of modeling power generation of wind turbine.
w_{co}, w_{ci}, and w_r are cutout, cut-in, and rated speeds of the wind turbine, respectively.
$w(t)$ is the wind speed.

By using the electrical storage system, the excessive electricity generation is stored to be used during peak periods. Equation (12.26) presents the available electric power in the storage:

$$C_t^{\text{st},e} = C_{t-1}^{\text{st},e} + p_t^{\text{ch},e} \times \eta_{\text{ch}}^e - p_t^{\text{dis},e}/\eta_{\text{dis}}^e - p_t^{\text{loss},e} \tag{12.26}$$

where $C_t^{\text{st},e}$ is the available electrical energy in the storage and $p_t^{\text{loss},e}$ is the energy loss of the electrical storage.

Equation (12.27) limits the stored energy in the storage system:

$$\alpha_{\min}^e \times C_c^{\text{st},e} \leq C_t^{\text{st},e} \leq \alpha_{\max}^e \times C_c^{\text{st},e} \tag{12.27}$$

Electricity energy loss in the storage is modeled as function of available energy in the storage system which is provided by Eq. (12.28):

$$p_t^{\text{loss},e} = \alpha_{\text{loss}}^e \times C_t^{\text{st},e} \tag{12.28}$$

where α_{loss}^e is the coefficient for modeling the energy loss in the electrical storage.

Equations (12.29) and (12.30) are used to limit charging and discharging power of the storage system, respectively. Equation (12.31) is implemented to prevent charging and discharging the storage at the same time:

$$\frac{\alpha_{\min}^e \times C_c^{\text{st},e} \times I_t^{\text{ch},e}}{\eta_{\text{ch}}^e} \leq p_t^{\text{ch},e} \leq \frac{\alpha_{\max}^e \times C_c^{\text{st},e} \times I_t^{\text{ch},e}}{\eta_{\text{ch}}^e} \tag{12.29}$$

$$\alpha_{\min}^e \times C_c^{\text{st},e} \times I_t^{\text{dis},e} \times \eta_{\text{dis}}^e \leq p_t^{\text{dis},e} \leq \alpha_{\max}^e \times C_c^{\text{st},e} \times I_t^{\text{dis},e} \times \eta_{\text{dis}}^e \tag{12.30}$$

$$I_t^{\text{ch},e} + I_t^{\text{dis},e} \leq 1 \tag{12.31}$$

where $I_t^{\text{ch},e}/I_t^{\text{dis},e}$ is the binary variable for charging and discharging state of electrical storage.

12.2.5 Modeling of Time-of-Use Rate of DRP

In order to deal with peak load demand periods which are experienced only few hours during a year, many efforts have been done [8]. To cope with this problem, demand response programs are considered as a feasible solution which is detailed in previous sections. In this chapter, TOU pricing is utilized to enhance both economic and environmental operations of the system [9, 10]. Equation (12.32) provides the mathematical formulation of the TOU rate of DRP:

$$p_t^{\mathrm{el,DRP}} = p_t^{\mathrm{el}} + p_t^{\mathrm{shup},e} - p_t^{\mathrm{shdo},e} \tag{12.32}$$

where $p_t^{\mathrm{el,DRP}}$ is the electrical demand under DRP.

Equations (12.33) and (12.34) limit the increased and decreased load in DRP, respectively:

$$0 \le p_t^{\mathrm{shup},e} \le LPF^{\mathrm{shup},e} \times p_t^l \times I_t^{\mathrm{shup},e} \tag{12.33}$$

$$0 \le p_t^{\mathrm{shdo},e} \le LPF^{\mathrm{shdo},e} \times p_t^l \times I_t^{\mathrm{shdo},e} \tag{12.34}$$

where $I_t^{\mathrm{shup},e}/I_t^{\mathrm{shdo},e}$ is the binary variable for increased/decreased load.

Increase or decrease of load under DRP at the same time is prevented by implementing Eq. (12.35):

$$I_t^{\mathrm{shup},e} + I_t^{\mathrm{shdo},e} \le 1 \tag{12.35}$$

Finally, increased and decreased loads should be equal at the end of the day which is expressed by Eq. (12.36):

$$\sum_t^H p_t^{\mathrm{shup},e} = \sum_t^H p_t^{\mathrm{shdo},e} \tag{12.36}$$

12.2.6 Other Technical Constraints

Equation (12.37) provides balance between the demand and procured water:

$$\mathrm{wa}_t^l = \mathrm{wa}_t^{\mathrm{net}} \tag{12.37}$$

Equation (12.38) states that the imported gas from the gas network should be equal to the consumed gas by residential section, boiler, and CHP unit:

$$g_t^{\text{net}} = g_t^B + g_t^{\text{CHP}} + g_t^l \tag{12.38}$$

where g_t^{net} is the total purchased gas from gas network.

In this chapter, it is assumed that the water and gas networks have a rated capacity, which means purchased water and gas cannot exceed the rated capacities. Equations (12.39) and (12.40) provide limitations of imported amount of water and gas carriers from the grid:

$$\text{wa}_{\min} \leq \text{wa}_t^{\text{net}} \leq \text{wa}_{\max} \tag{12.39}$$

$$g_{\min}^{\text{net}} \leq g_t^{\text{net}} \leq g_{\max}^{\text{net}} \tag{12.40}$$

Equations (12.41) and (12.42) limit the electrical and thermal generation of the CHP unit, respectively:

$$\eta_{\text{ge}}^{\text{CHP}} \times g_t^{\text{CHP}} \leq p_c^{\text{CHP}} \tag{12.41}$$

$$\eta_{\text{gh}}^B \times g_t^B \leq p_c^B \tag{12.42}$$

12.3 Modeling of ε-Constraint and Max-Min Fuzzy Satisfying Methods

In order to solve the developed multi-objective model, the ε-constraint and max-min fuzzy satisfying methods which are used to solve the multi-objective model are introduced [2, 11, 12] in this section.

In the ε-constraint method, one of the objective functions is considered as the main objective function, and the other is considered as a constraint. Therefore, the problem is turned to a single-objective minimization problem as Eq. (12.43):

$$\begin{aligned} &\text{OF} = \min(\text{f}_1) \\ &\text{s.t.} \\ &\begin{cases} \text{f}_2 \leq \varepsilon \\ \text{Equal \& unequal equations} \end{cases} \end{aligned} \tag{12.43}$$

The Pareto front solutions are generated by changing the value of ε from $\text{f}_2^{\min}$ up to $\text{f}_2^{\max}$ in the main objective function of the multi-objective model.

When the problem is solved using the ε-constraint method for both objectives, the max-min fuzzy technique is employed to normalize both of the conflicting objective functions in Eq. (12.44):

$$\mu_k^n = \begin{cases} 1 & f_k^n \leq f_k^{\min} \\ \dfrac{f_k^{\max} - f_k^n}{f_k^{\max} - f_k^{\min}} & f_k^{\min} \leq f_k^n \leq f_k^{\max} \\ 0 & f_k^n \geq f_k^{\max} \end{cases} \tag{12.44}$$

where μ_k^n is the optimal level of nth solution of objective function k,$f_k^{\max}$ and$f_k^{\min}$ are the maximum and minimum values of objective function k, respectively.

After the process of normalization at each iteration, a comparison between per unit values of each objective function is carried out, and the minimum values are selected (12.45):

$$\mu^n = \min\left(\mu_1^n, \ldots, \mu_N^n\right); \quad \forall \mathrm{n} = 1, \ldots, \mathrm{N}_P \tag{12.45}$$

In order to get the best trade-off solution between the two conflicting objectives, the maximum value of the minimums is chosen (12.46):

$$\mu^{\max} = \max\left(\mu^1, \ldots, \mu^{\mathrm{Np}}\right) \tag{12.46}$$

12.4 Numerical Results

In this section, information of a sample energy hub system is provided. Also, obtained results of the ε-constraint is presented and discussed.

12.4.1 Input Data

Introduced energy hub model in previous sections including several types of energy resources is considered to supply different energy demands of a large consumer. Figure 12.2 provides the different load demands of the large consumer including electrical, water, gas, and heat [7].

The electricity price in the market is illustrated in Fig. 12.3. In addition, Fig. 12.4 presents the wind speed which is used to model the power output of the wind turbine. Table 12.1 presents the gas and water prices. Note that the gas is imported for operation of boiler and CHP units as well as operation costs of electrical and thermal storages, wind turbine, and DRP. Finally, required information of energy hub devices is provided in Table 12.2. The model is formulated as MILP and solved using GAMS optimization software [13] under CPLEX 11.0 [14].

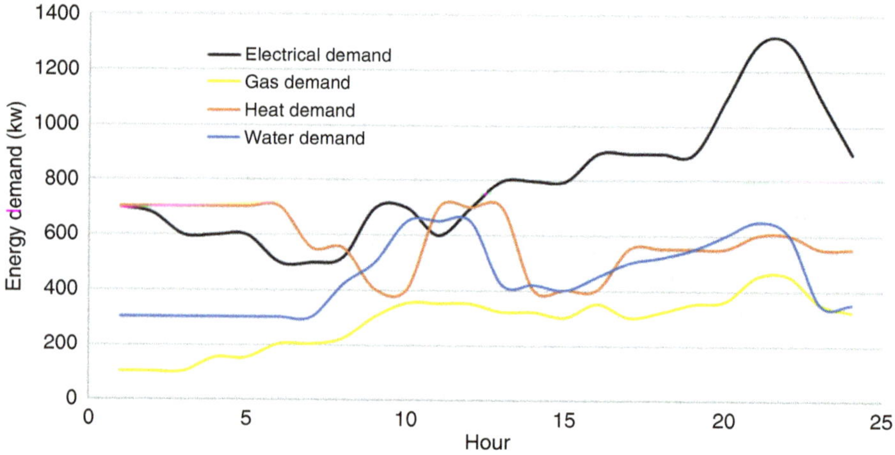

Fig. 12.2 Energy demand

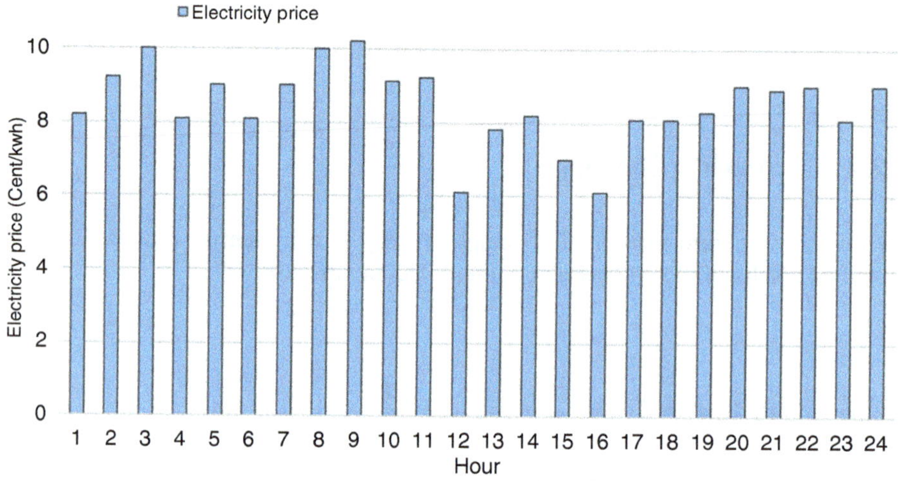

Fig. 12.3 Electricity power price in the market

12.4.2 Simulation Results

In this section, the results of the proposed model are provided within two case studies as follows:

Case 1: Optimal scheduling of the energy hub system without DRP
Case 2: Optimal scheduling of the energy hub system with DRP

In the first case, the problem is solved without implementing TOU rate of DRP. Figure 12.5 provides the solutions of the proposed multi-objective problem, which is

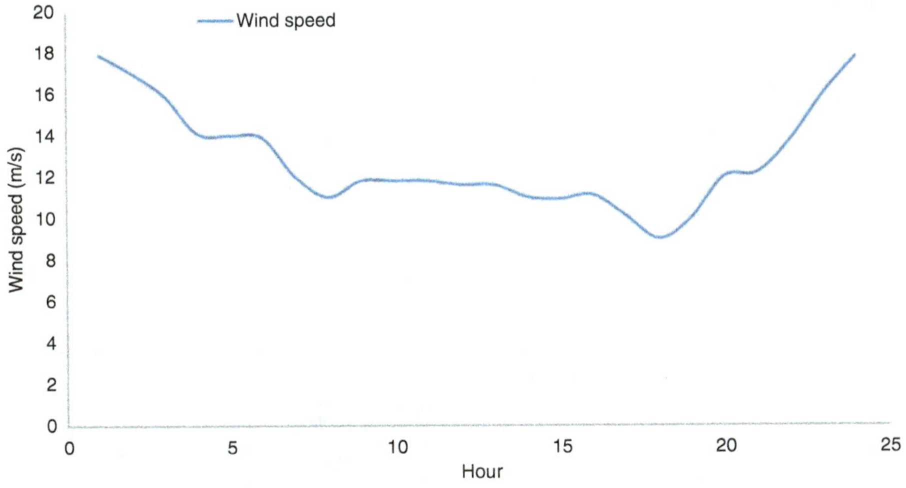

Fig. 12.4 Wind speed at each hour

Table 12.1 Prices of the hub energy system

Parameter	Value	Unit
λ^{g}	6	Cent/kWh
λ^{wa}	4	Cent/kWh
λ^{wi}	0	Cent/kWh
λ_{s}^{e}	2	Cent/kWh
λ_{s}^{h}	2	Cent/kWh
λ^{DR}	2	Cent/kWh

solved using the ε-constraint method in 20 iterations. Note that in Fig. 12.5, the selected solution is presented in yellow and green colors for without and with DRP cases, respectively. Table 12.3 provides detailed information about the obtained results of with and without DRP cases. Note the first row in the Table 12.3 provides the minimum operating cost in both cases and the last row presents the minimum emission.

In the first case, solution #12 is selected as the best compromising solution while in the second case, the solution #11 is chosen based on the fuzzy satisfying decision-making method.

In the without DRP case, total operating cost and emission is obtained as $2719.65 and 10,382.2 kg/day, respectively.

After implementing TOU rate of DRP in the second case, solution #11 is selected as the best compromise solution using the max-min fuzzy satisfying method. In this case, the results are obtained as $2681.138 and 10,347.6 kg/day for operating cost and emission, respectively.

By comparing obtained results in two cases, it appeared that by implementing TOU-DRP, the total operating cost is reduced about 1.41%. Also, the TOU-DRP has

Table 12.2 Information of utilized devices in the energy hub system

#	Unit	Value	#	Unit	Value	#	Unit	Value
Electrical storage parameter			**Thermal storage parameter**			**Wind turbine parameter**		
α^{e}_{min}	–	0.05	α^{h}_{min}	–	0.05	A^{WIND}	–	0.96
α^{e}_{max}	–	0.9	α^{h}_{max}	–	0.9	x,y,z	–	0.07,0.01,0.03
α^{e}_{loss}	–	0.2	α^{h}_{loss}	–	0.2	w_r	m/s	10
η^{e}_{ch}	%	90	η^{h}_{ch}	%	90	w_{ci}	m/s	4
η^{e}_{dis}	%	90	η^{h}_{dis}	%	90	w_{co}	m/s	22
$C^{st,e}_{c}$	kW	300	$C^{st,h}_{c}$	kW	200	p_r	kW	400
CHP parameter			**Upstream network parameter**			**Boiler, gas, and water network parameters**		
η^{CHP}_{ge}	%	40	A^{NET}	–	0.99	η^{B}_{gh}	%	85
η^{CHP}_{gh}	%	35	p^{e}_{max}	kW	1000	p^{B}_{c}	kW	800
A^{CHP}	–	0.96	p^{e}_{min}	kW	0	g^{net}_{max}	kW	1800
p^{CHP}_{c}	kW	800	p^{T}_{c}	kW	800	wa_{max}	kW	1000
Boiler emission			**Upstream network emission**			**CHP emission**		
EF^{B}_{CO}	kg/kWh	0.37	EF^{Net}_{CO}	kg/kWh	0.368	EF^{CHP}_{CO}	kg/kWh	0.37
EF^{B}_{SO}	kg/kWh	0.000003	EF^{Net}_{SO}	kg/kWh	0.0002	EF^{CHP}_{SO}	kg/kWh	0.000003
EF^{B}_{NO}	kg/kWh	0.00009	EF^{Net}_{NO}	kg/kWh	0.0008	EF^{CHP}_{NO}	kg/kWh	0.00009

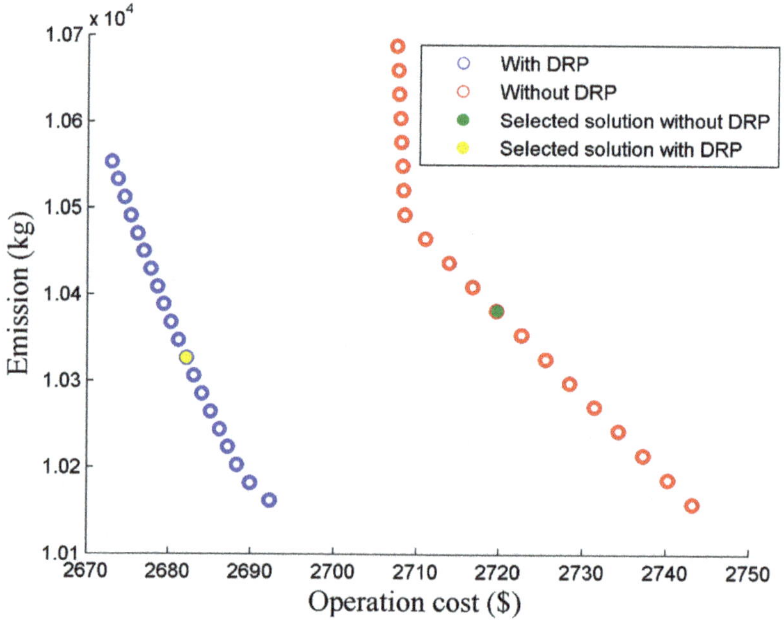

Fig. 12.5 Pareto front and selected solutions in with and without DRP cases

Table 12.3 Pareto front solutions for without and with DRP cases

#	Without DRP					#	With DRP				
	Cost ($)	Emission (kg)	$\Phi_1\,(p.\,u.)$	$\Phi_2\,(p.\,u.)$	$\min(\Phi_1, \Phi_2)$		Cost ($)	Emission (kg)	$\Phi_1\,(p.\,u.)$	$\Phi_2\,(p.\,u.)$	$\min(\Phi_1, \Phi_2)$
1	2707.335	10,688.5	1	0	0	1	2672.913	10,553.53	1	0	0
2	2707.486	10,660.65	0.9958	0.0526	0.0526	2	2673.71	10,532.94	0.9587	0.0526	0.0526
3	2707.636	10,632.81	0.9916	0.1053	0.1053	3	2674.507	10,512.34	0.9175	0.1053	0.1053
4	2707.787	10,604.96	0.9874	0.1579	0.1579	4	2675.303	10,491.75	0.8762	0.1579	0.1579
5	2707.938	10,577.12	0.9831	0.2105	0.2105	5	2676.1	10,471.16	0.8350	0.2105	0.2105
6	2708.089	10,549.27	0.9789	0.2632	0.2632	6	2676.897	10,450.56	0.7937	0.2632	0.2632
7	2708.239	10,521.43	0.9747	0.3158	0.3158	7	2677.693	10,429.97	0.7524	0.3158	0.3158
8	2708.39	10,493.58	0.9705	0.3684	0.3684	8	2678.49	10,409.38	0.7112	0.3684	0.3684
9	2710.859	10,465.74	0.9014	0.4211	0.4211	9	2679.287	10,388.78	0.6699	0.4211	0.4211
10	2713.79	10,437.89	0.8195	0.4737	0.4737	10	2680.204	10,368.19	0.6224	0.4737	0.4737
11	2716.72	10,410.05	0.7375	0.5263	0.5263	**11**	**2681.138**	**10,347.6**	**0.5741**	**0.5263**	**0.5263**
12	**2719.65**	**10,382.2**	**0.6556**	**0.5789**	**0.5789**	12	2682.072	10,327	0.5257	0.5789	0.5257
13	2722.58	10,354.36	0.5736	0.6316	0.5736	13	2683.014	10,306.41	0.4769	0.6316	0.4769
14	2725.51	10,326.51	0.4917	0.6842	0.4917	14	2684.029	10,285.82	0.4243	0.6842	0.4243
15	2728.44	10,298.67	0.4097	0.7368	0.4097	15	2685.1	10,265.22	0.3689	0.7368	0.3689
16	2731.37	10,270.82	0.3278	0.7895	0.3278	16	2686.171	10,244.63	0.3134	0.7895	0.3134
17	2734.3	10,242.98	0.2458	0.8421	0.2458	17	2687.242	10,224.04	0.2580	0.8421	0.2580
18	2737.23	10,215.13	0.1639	0.8947	0.1639	18	2688.312	10,203.44	0.2025	0.8947	0.2025
19	2740.16	10,187.29	0.0819	0.9474	0.0819	19	2689.899	10,182.85	0.1203	0.9474	0.1203
20	2743.09	10,159.45	0	1	0	20	2692.223	10,162.26	0	1	0

Bold values are the best compromising solutions based on the fuzzy satisfying decision-making method

Table 12.4 Comparison results of cases 1 and 2

Parameters	Without DRP	With DRP
Total cost ($)	2719.65	2681.138
Total emission (kg)	10,382.2	10,347.6
Cost reduction (%)	0	1.41
Emission reduction (kg/day)	0	34.6

reduced emission about 34.6 kg/day in the second case. A summary of obtained results of both without and with DRP cases is presented in Table 12.4.

The minimum operating cost of the energy hub is reported as $2707.335 and $2672.913 for without and with DRP cases in solution #1, respectively. Note that the maximum emission reported in this solution is equal to 10,688.5 and 10,553.53 kg/day for without and with DRP cases. The results show 1.27% and 1.26% for total cost and emission due to the implementation of the DRP.

The minimum amount of emission which is obtained in the solution #20 is reported as 10,159.45 and 10,162.26 kg/day for without and with DRP. Note that the maximum operating cost reported in this solution is equal to $2743.09 and $2692.223 for without and with DRP cases.

12.4.3 Performance of the Devices

In this section, the performances of different devices are compared in both cases. The electrical load demand for without and with DRP cases is presented in Fig. 12.6. As was expected, DRP shifts some part of the load form peak periods to off-peak periods.

Procured power from the upstream grid to supply the electrical load demand of the hub energy system is shown in Fig. 12.7 for both with and without DRP cases. According to Fig. 12.7, purchased power in peak periods has been reduced, while in off-peak periods, it has been increased which has led to less operation cost of hub energy system, in with DRP case.

The electrical generation of CHP unit is presented in Fig. 12.8. Total electrical generation of CHP is equal to 1739.1 and 1316.3 kWh for without and with DRP, respectively. The generation of CHP units is reduced by about 24.31% due to the implementation of RTP-DRP.

The electrical storage is implemented to supply some part of the required load demand. Stored energy in the storage is illustrated in Fig. 12.9. In addition, Fig. 12.10 presents the charging and discharging states of the storage during time horizon of the study for without and with DRP case. According to Figures 12.9 and 12.10, the initial stored energy in the storage is equal to 30 kW, which is increased to 140.8 kW by charging 110.8 kW at hour 2 in with DRP case. It is obvious that the initial load in without DRP case unit is at hour 6, which is charged about 240 kW. The charging and discharging process is repeated during the 24 h to supply the electrical load demand of the energy hub.

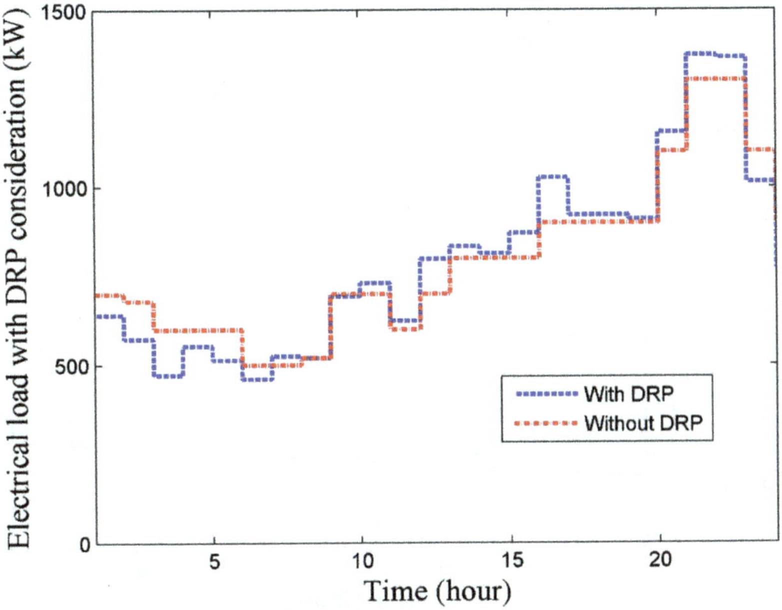

Fig. 12.6 Electrical load in the presence of DRP

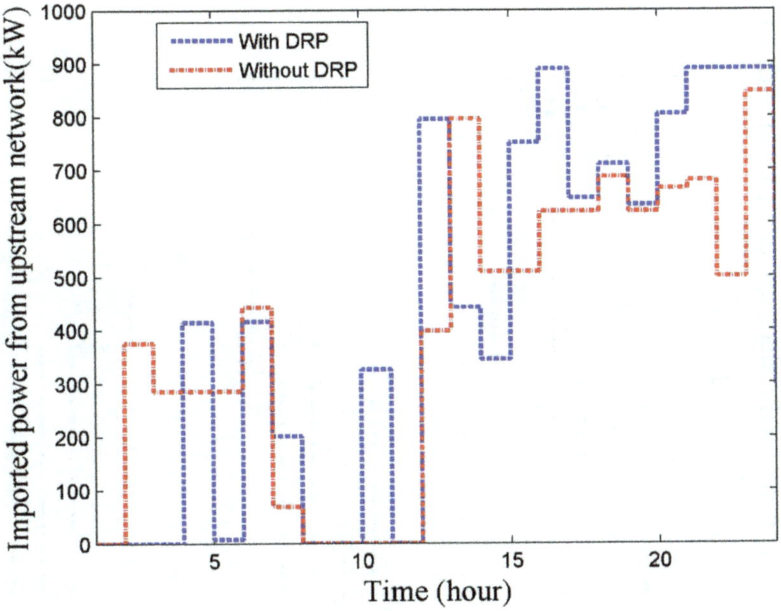

Fig. 12.7 Purchased power form upstream grid

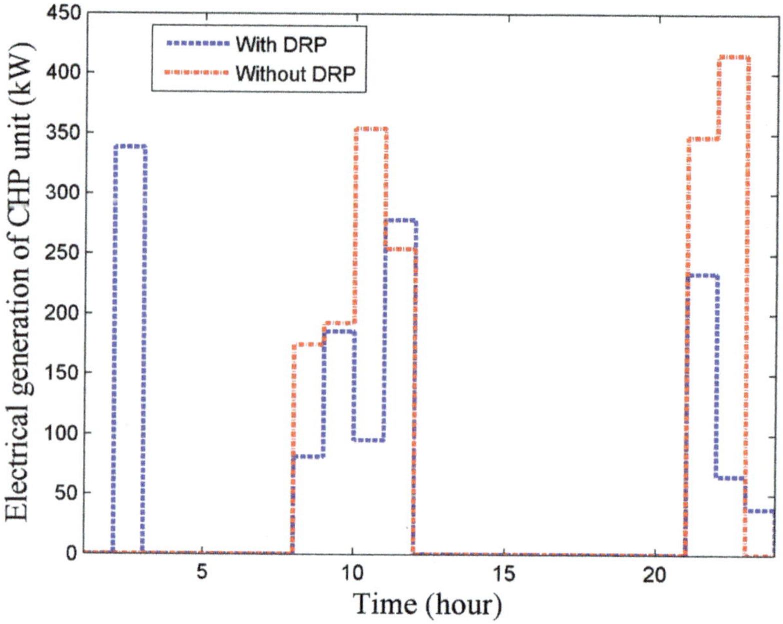

Fig. 12.8 Electrical generation of CHP unit

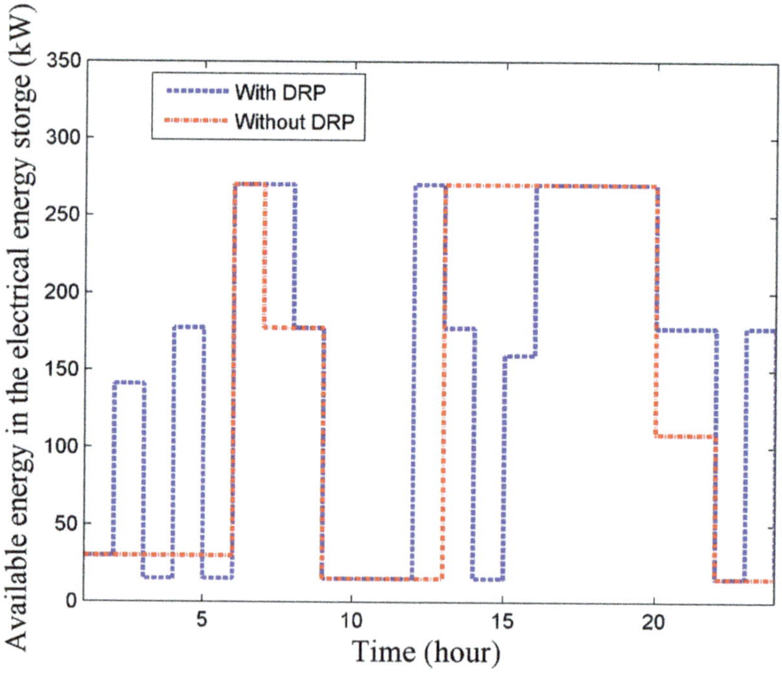

Fig. 12.9 Available energy in the electrical storage system

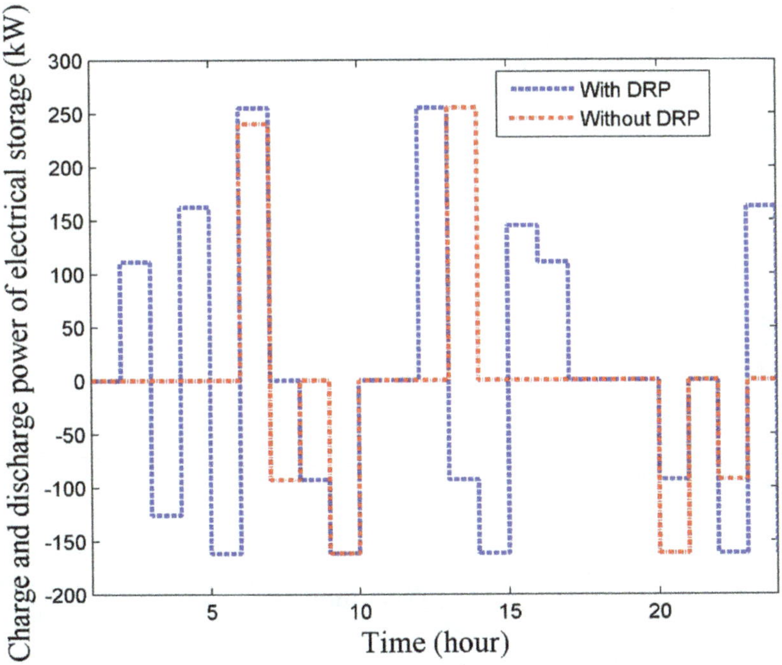

Fig. 12.10 Charging and discharging states of the electrical storage system

In order to supply the heat demand of the energy hub, various devices such as CHP, boiler, and thermal storage have been utilized. Heat generation of the CHP units is presented in Fig. 12.11. Note that the thermal and electrical generations of the CHP unit are dependent to each hour. Therefore, heat generation of the CHP is changed in accordance with the heat generation of the CHP unit. As it can be seen, the generated heat by the CHP is reduced in with DRP case in which the total generated heat is recorded as 1521.6 and 1151.7 kW for without and with DRP cases.

The generated heat by the boiler is illustrated in Fig. 12.12. In the second case, the heat generation of the boiler is increased to supply the heat load demand.

Finally, available heat energy in the heat storage is depicted in Fig. 12.13 for both without and with DRP cases. In addition, the charging and discharging of the thermal storage is depicted in Fig. 12.14. Total amount of the stored energy in the thermal storage is equal to 544 and 592 kWh for with and without DRP cases. It is obvious that the initial stored energy in the thermal storage is equal to 20 kW which is increased about 26 kW at hour 2 in with DRP case, while in without DRP case, the thermal storage discharged at hour 7 is about 9 kW. The charging and discharging process is continued in time horizon of the study to supply the heat load demand of the energy hub.

Figure 12.15 presents the purchased gas from the upstream gas grid. As said before, purchased gas is used to feed the CHP and boiler to generate electrical and

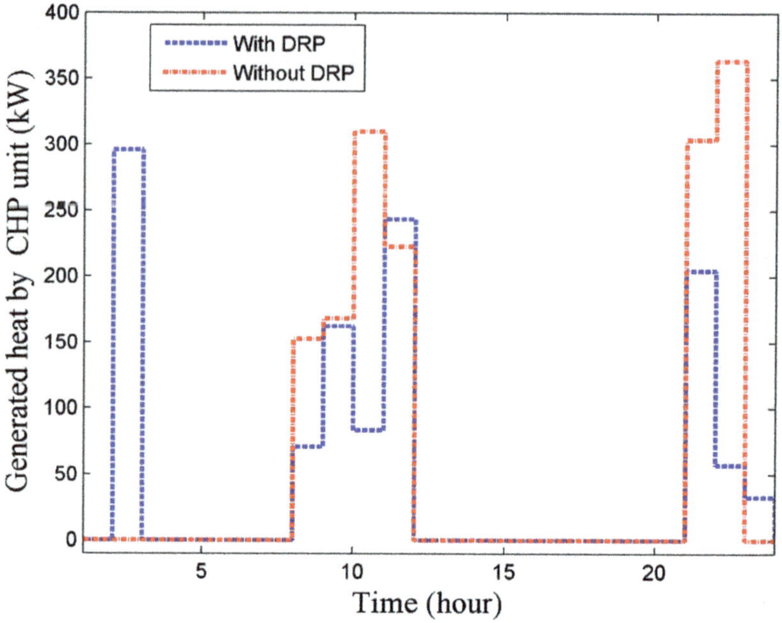

Fig. 12.11 Heat generation of the CHP unit

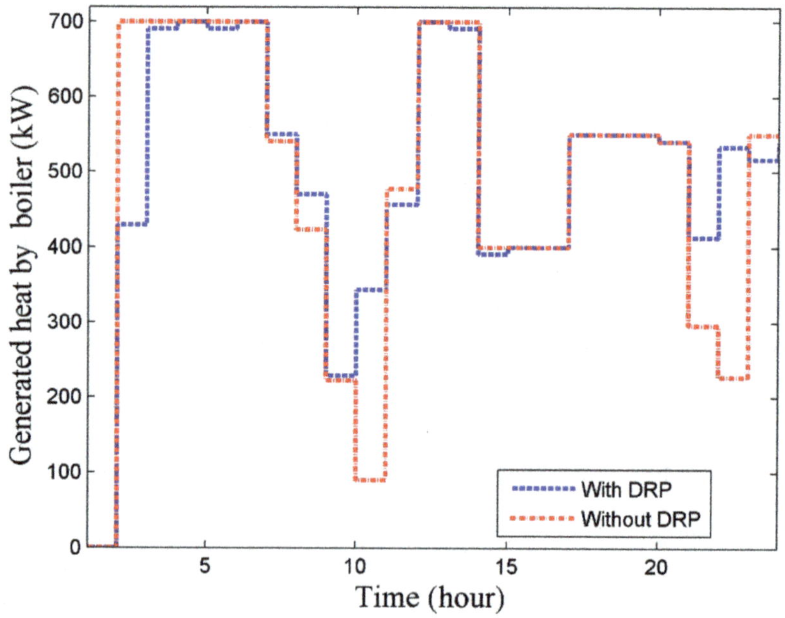

Fig. 12.12 Heat generation of the boiler unit

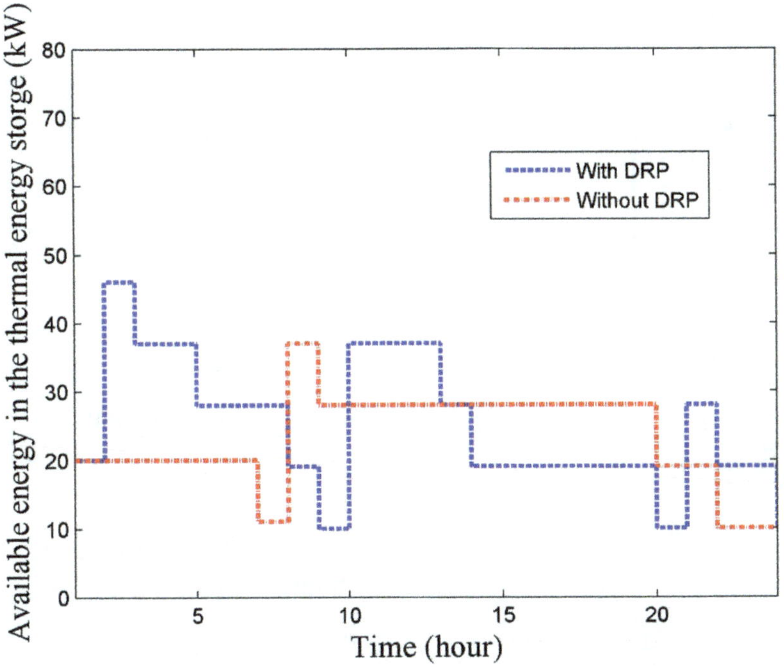

Fig. 12.13 Heat generation of the boiler unit

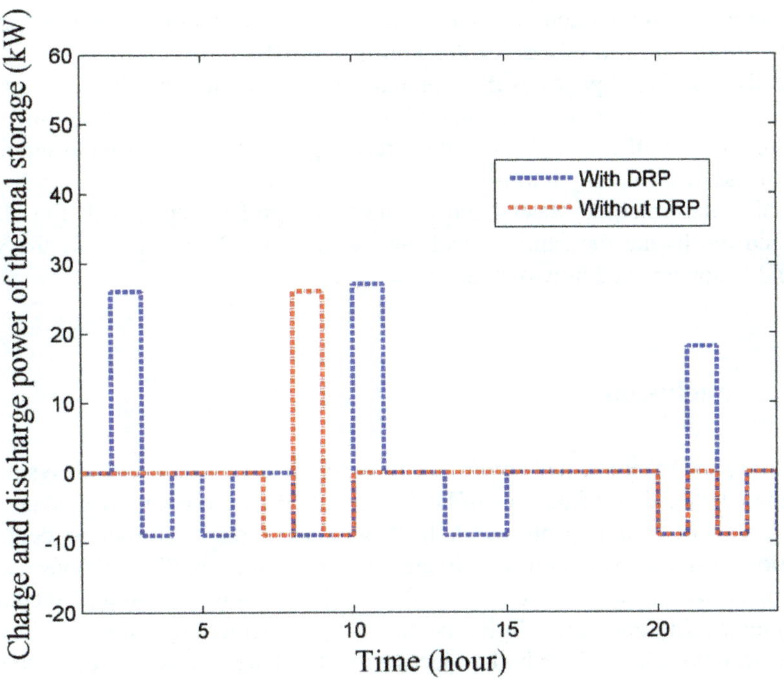

Fig. 12.14 Charging and discharging of the thermal storage system

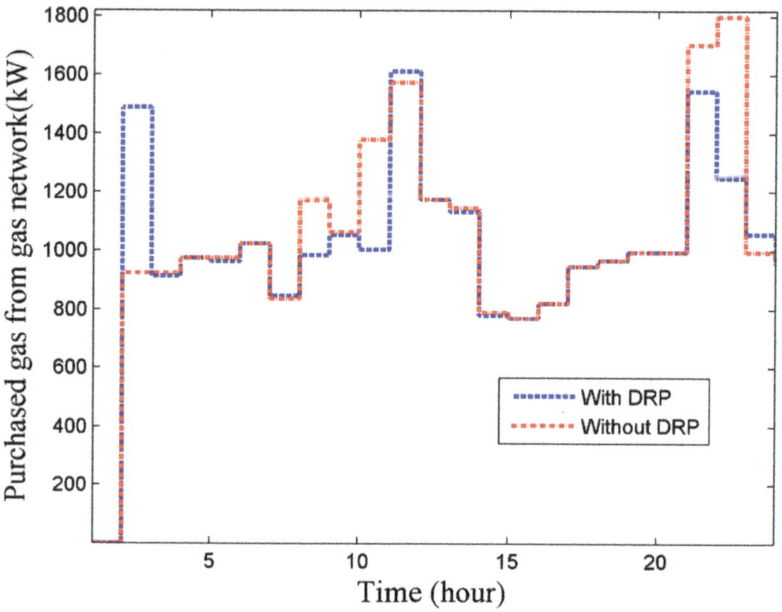

Fig. 12.15 Purchased gas from the upstream gas grid

heat energy. In addition, some part of the procured gas is utilized to meet the gas demand of the energy hub. Figure 12.16 presents the purchased gas for the CHP unit. Note that due to the reduction of electrical and heat generation of the CHP unit in with DRP case, purchased gas for this unit is reduced.

Finally, Fig. 12.17 provides the inputted gas to the boiler unit. It is obvious that purchased gas for the boiler is increased because the heat generation of this unit is increased in with DRP case. Note that the rest of purchased gas is used to supply the gas demand of the energy hub in the residential section.

Finally, the procured water from the upstream grid is depicted in Fig. 12.18. Obtained results are the same in with and without DRP case, because the water demand is not changed in two cases.

12.5 Conclusion

In this chapter, economic and environmental operations of an energy hub system are studied under the time-of-use rate of DRP to supply the various energy demands of a large consumer including electrical, heat, gas, and water. In order to deal with operation cost and emission which are known as the conflicting objectives, a multi-objective model is developed and solved using the ε-constraint method in 20 iterations. The trade-off solution is chosen by implementing the max-min fuzzy satisfying approach in which both objective functions are satisfied. In order to show

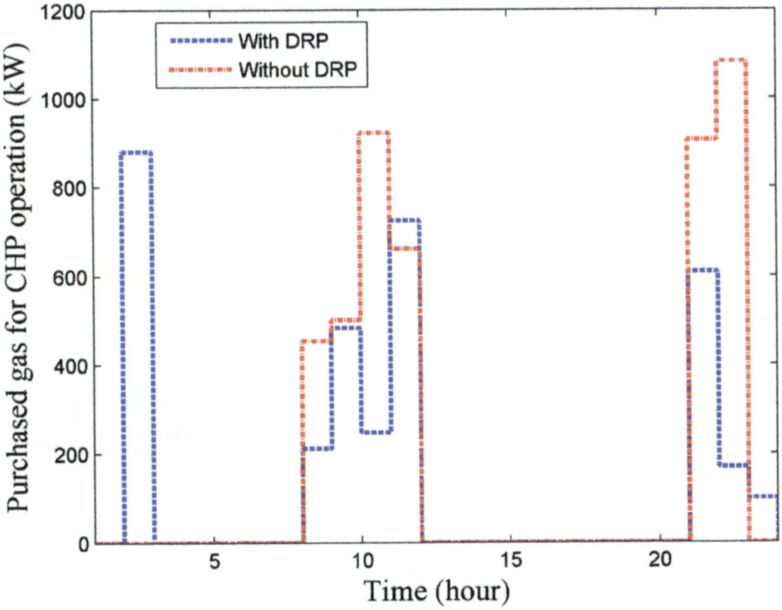

Fig. 12.16 Purchased gas for the CHP unit

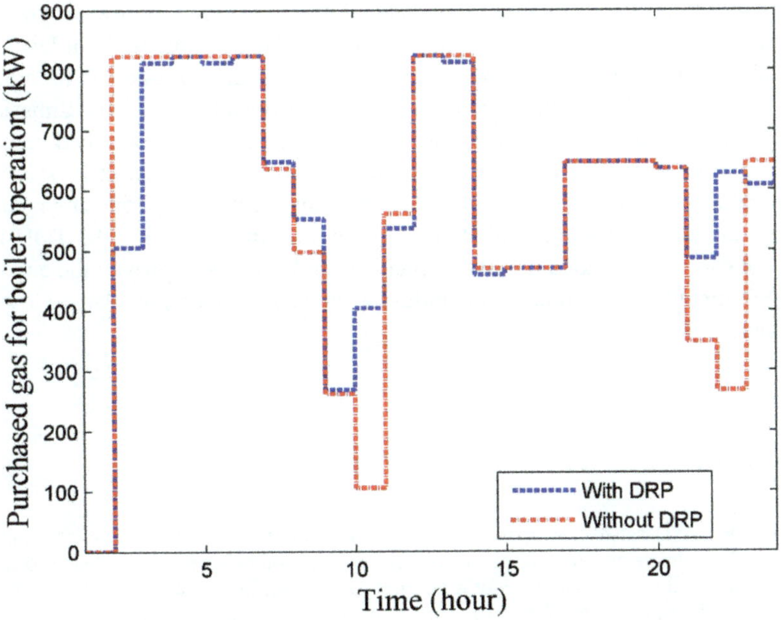

Fig. 12.17 Purchased gas for the boiler unit

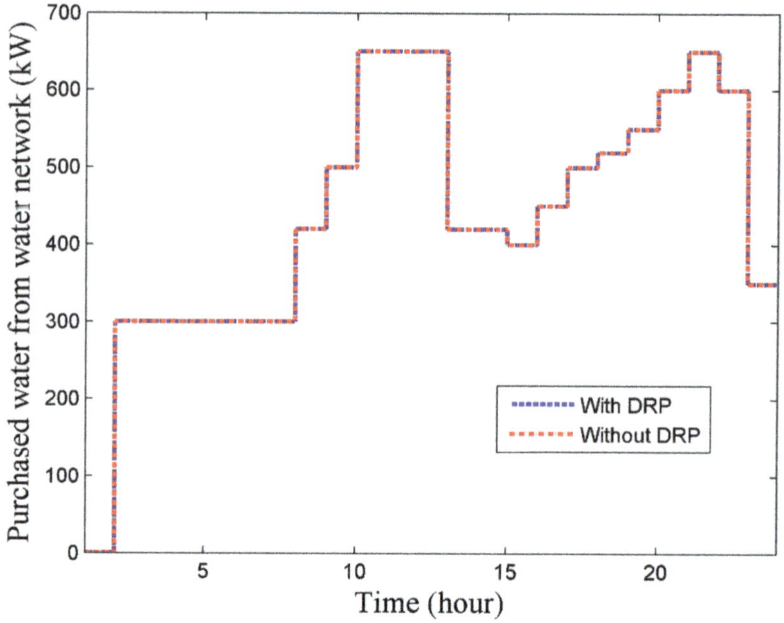

Fig. 12.18 Purchased water for the upstream

the impact of the time-of-use pricing rate of the demand response program, the problem is solved in two cases as with and without DRP. In the without DRP case, the total operation cost is obtained as $2719.65, and the emission is recorded as 10,382.2 kg/day. Also, in the second case, the total operation cost and emission of hub energy system are $2681.138 and 10,347.6, respectively. Obtained results in two cases show that the total operating cost of the hub is reduced about 1.41%, which is in accordance with our economic goals. In addition, by comparing the results, it says that the emission is reduced about 34.6 kg after implementing the TOU-DRP. Finally, it can be concluded that the implementing DRP has improved the economic and environmental operations of the multi-carrier energy system of a large electricity consumers.

References

1. M. Hamian, A. Darvishan, M. Hosseinzadeh, M.J. Lariche, N. Ghadimi, A. Nouri, A framework to expedite joint energy-reserve payment cost minimization using a custom-designed method based on mixed integer genetic algorithm. Eng. Appl. Artif. Intell. **72**, 203–212 (2018)
2. M. Majidi, S. Nojavan, N. Nourani Esfetanaj, A. Najafi-Ghalelou, K. Zare, A multi-objective model for optimal operation of a battery/PV/fuel cell/grid hybrid energy system using weighted sum technique and fuzzy satisfying approach considering responsible load management. Sol. Energy **144**, 79–89 (2017)

3. S. Nojavan, M. Majidi, A. Najafi-Ghalelou, M. Ghahramani, K. Zare, A cost-emission model for fuel cell/PV/battery hybrid energy system in the presence of demand response program: ε-constraint method and fuzzy satisfying approach. Energy Convers. Manag. **138**, 383–392 (2017)
4. A. Bal, S.I. Satoglu, A goal programming model for sustainable reverse logistics operations planning and an application. J. Clean. Prod. **201**, 1081–1091 (2018)
5. H. Khodaei, M. Hajiali, A. Darvishan, M. Sepehr, N. Ghadimi, Fuzzy-based heat and power hub models for cost-emission operation of an industrial consumer using compromise programming. Appl. Therm. Eng. **137**, 395–405 (2018)
6. G. Aghajani, N. Ghadimi, Multi-objective energy management in a micro-grid. Energy Rep. **4**, 218–225 (2018)
7. S. Pazouki, M.-R. Haghifam, A. Moser, Uncertainty modeling in optimal operation of energy hub in presence of wind, storage and demand response. Int. J. Electr. Power Energy Syst. **61**, 335–345 (2014)
8. M. Majidi, S. Nojavan, K. Zare, Optimal stochastic short-term thermal and electrical operation of fuel cell/photovoltaic/battery/grid hybrid energy system in the presence of demand response program. Energy Convers. Manag. **144**, 132–142 (2017)
9. S. Nojavan, H. Qesmati, K. Zare, H. Seyyedi, Large consumer electricity acquisition considering time-of-use rates demand response programs. Arab. J. Sci. Eng. **39**(12), 8913–8923 (2014)
10. S. Nojavan, H. Ghesmati, K. Zare, Robust optimal offering strategy of large consumer using IGDT considering demand response programs. Electr. Power Syst. Res. **130**, 46–58 (2016)
11. S. Nojavan, M. Majidi, N.N. Esfetanaj, An efficient cost-reliability optimization model for optimal siting and sizing of energy storage system in a microgrid in the presence of responsible load management. Energy **139**, 89–97 (2017)
12. A.S.M. Rabiee, S. Jalilzadeh, B. Mohammadi-Ivatloo, S. Nojavan, Probabilistic multi objective optimal reactive power dispatch considering load uncertainties using Monte Carlo simulations. J. Oper. Autom. Power Eng. **3**(1), 83–93 (2015)
13. A. Brooke, D. Kendrick, A. Meeraus, R. Raman, R.E. Rosenthal, *GAMS A User's Guide Introduction 1* (GAMS Development Corporation, Washington, DC, 1998)
14. CPLEX 12. [Online]. Available: https://www.gams.com/latest/docs/S_CPLEX.html. Accessed 15 Jul 2018

Index

S. Nojavan et al. (eds.), *Robust Energy Procurement of Large Electricity Consumers*,
https://doi.org/10.1007/978-3-030-03229-6

CPSIA information can be obtained
at www.ICGtesting.com
Printed in the USA
LVHW050101171218
600703LV00005B/331/P

9 783030 032289